Petroleum Geochemistry

Petroleum Geochemistry

Chinmay Chatterjee

RANDOM

RANDOM PUBLICATIONS

NEW DELHI (INDIA)

Petroleum Geochemistry

ISBN 978-93-5111-419-2

Published in 2014 in India by

RANDOM PUBLICATIONS

4376-A/4B, Gali Murari Lal, Ansari Road
New Delhi-110 002
Phone : +91-11-43580356, +91-11-23289044
e-mail: randomexports@gmail.com, sales@randompublications.com, info@randompublications.com

Reprinted 2026

Type Setting by: Friends Media, Delhi-110089

Preface

Petroleum geochemistry is the application of chemical principles to the study of the origin, migration, accumulation, and alteration of petroleum...(John M. Hunt, 1979). Petroleum is generally considered oil and natural gases having various compounds composed of primarily hydrogen and carbon. They are usually generated from the decomposition and/or thermal maturation of organic matter. The organic matter originated from plants and algae. The organic matter is deposited after the death of the plant in sediments, where after considerable time, heat, and pressure the compounds in the plants and algae are altered to oil, gas, and kerogen. Kerogen can be thought of as the remaining solid material of the plant. The sediment - usually clay and/or calcareous (lime) ooze, hardens during this alteration process into rock i.e. shale and/or limestone. The shale or limestone rock containing the organic matter is called the source rock because it is the source, having generated the petroleum.

In our quest to find more oil and gas, we use numerous petroleum geochemical techniques to; identify source rocks and determine the amount, type, and maturation level of the organic matter; evaluate the potential timing of petroleum migration from the source rock; assess the potential migration pathways; and correlate petroleum compounds found in reservoirs, leaks, and surface seeps to find new pools of petroleum. Some of the petroleum geochemical techniques used are: stable isotopes; hydrocarbon analysis (fingerprinting) of specific organic compounds (biological organic markers) with gas chromatography (GC) and gas chromatography-mass spectrometry (GC-MS); maturity indicators like vitrinite reflectance; laboratory pyrolysis and analysis; kerogen typing; and other more exotic techniques.

The book is primarily aimed at geologists and geophysicists with little previous knowledge of geochemistry, although it also includes advanced concepts of geochemistry including biomarker analysis. This book focuses on the dynamic petroleum system concept, exploration geochemistry of conventional and unconventional petroleum, and reservoir geochemistry. The book is designed for exploration, production, and development geologists. The book is intended to provide an understanding of geochemistry that can be applied by explorationists in prospect appraisal and exploration decision making.

I thank all members of my team who have helped in the preparation of the book. My special thanks go to "Random Publication" who have published the book.

—Chinmay Chatterjee

Contents

1

Introduction

The field of geochemistry involves study of the chemical composition of the Earth and other planets, chemical processes and reactions that govern the composition of rocks, water, and soils, and the cycles of matter and energy that transport the Earth's chemical components in time and space, and their interaction with the hydrosphere and the atmosphere.

Some subsets of geochemistry are:

Isotope Geochemistry: Determination of the relative and absolute concentrations of the elements and their isotopes in the earth and on earth's surface. Examination of the distribution and movements of elements in different parts of the earth (crust, mantle, hydrosphere etc.) and in minerals with the goal to determine the underlying system of distribution and movement.

Cosmochemistry: Analysis of the distribution of elements and their isotopes in the cosmos.

Biogeochemistry: Field of study focusing on the effect of life on the chemistry of the earth.

Organic Geochemistry: A study of the role of processes and compounds that are derived from living or once-living organisms.

Aqueous Geochemistry: Understanding the role of various elements in watersheds, including copper, sulphur, mercury, and how elemental fluxes are exchanged through atmospheric-terrestrial-aquatic interactions. Regional, environmental and exploration geochemistry: Applications to environmental, hydrological and mineral exploration studies. Victor Goldschmidt is considered by most to be the father of modern geochemistry and the ideas of the subject were formed by him in a series of publications from 1922 under the title

'Geochemische Verteilungsgesetze der Elemente' (geochemical laws of distribution of the elements).

Chemical Characteristics

The more common rock constituents are nearly all oxides; chlorides, sulphides and fluorides are the only important exceptions to this and their total amount in any rock is usually much less than 1%. F. W. Clarke has calculated that a little more than 47% of the Earth's crust consists of oxygen. It occurs principally in combination as oxides, of which the chief are silica, alumina, iron oxides, and various carbonates (calcium carbonate, magnesium carbonate, sodium carbonate, and potassium carbonate). The silica functions principally as an acid, forming silicates, and all the commonest minerals of igneous rocks are of this nature. From a computation based on 1672 analyses of numerous kinds of rocks Clarke arrived at the following as the average percentage composition: SiO_2=59.71, Al_2O_3=15.41, Fe_2O_3=2.63, FeO=3.52, MgO=4.36, CaO=4.90, Na_2O=3.55, K_2O=2.80, H_2O=1.52, TiO_2=0.60, P_2O_5=0.22, total 99.22%). All the other constituents occur only in very small quantities, usually much less than 1%.

These oxides combine in a haphazard way. For example, potash (potassium carbonate) and soda (sodium carbonate) combine to produce feldspars. In some cases they may take other forms, such as nepheline, leucite, and muscovite, but in the great majority of instances they are found as feldspar. Phosphoric acid with lime (calcium carbonate) forms apatite. Titanium dioxide with ferrous oxide gives rise to ilmenite. Part of the lime forms lime feldspar. Magnesium carbonate and iron oxides with silica crystallize as olivine or enstatite, or with alumina and lime form the complex ferro-magnesian silicates of which the pyroxenes, amphiboles, and biotites are the chief. Any excess of silica above what is required to neutralize the bases will separate out as quartz; excess of alumina crystallizes as corundum. These must be regarded only as general tendencies. It is possible, by rock analysis, to say approximately what minerals the rock contains, but there are numerous exceptions to any rule.

Mineral Constitution

Hence we may say that except in acid or siliceous rocks containing 66% of silica and over, quartz will not be abundant. In basic rocks (containing 20% of silica or less) it is rare and accidental. If magnesia and iron be above the average while silica is low, olivine may be expected; where silica is present in greater quantity over ferro-magnesian minerals, such as augite, hornblende, enstatite or biotite, occur rather than olivine. Unless potash is high and silica relatively low, leucite will not be present, for leucite does not occur with free quartz. Nepheline, likewise, is usually found in rocks with much soda and comparatively little silica. With high alkalis, soda-bearing pyroxenes and amphiboles may be present. The lower the percentage of silica and the alkalis, the greater is the prevalence of calcium feldspar as contracted with soda or potash feldspar. Clarke has calculated the relative abundance of the principal rock-forming minerals with the following results: apatite=0.6, titanium minerals=1.5, quartz=12.0, feldspars=59.5, biotite=3.8, hornblende and pyroxene=16.8, total=94.2%. This, however, can only be a rough approximation.

The other determining factor, namely the physical conditions attending consolidation, plays on the whole a smaller part, yet is by no means negligible, as a few instances will prove. Certain minerals are practically confined to deep-seated intrusive rocks, e.g., microcline, muscovite, diallage. Leucite is very rare in plutonic masses; many minerals have special peculiarities in microscopic character according to whether they crystallized in depth or near the surface, e.g., hypersthene, orthoclase, quartz.

There are some curious instances of rocks having the same chemical composition, but consisting of entirely different minerals, e.g., the hornblendite of Gran, in Norway, which contains only hornblende, has the same composition as some of the camptonites of the same locality that contain feldspar and hornblende of a different variety. In this connection we may repeat what has been said above about the corrosion of porphyritic minerals in igneous rocks. In rhyolites and trachytes, early crystals of hornblende and biotite may be found in great numbers partially converted into augite and magnetite. Hornblende and biotite were stable under the pressures and other conditions below the surface, but unstable at higher levels. In the ground-mass of these rocks, augite is almost universally present. But the plutonic representatives of the same magma, granite and syenite contain biotite and hornblende far more commonly than augite.

Acid, Intermediate and Basic Igneous Rocks

Those rocks that contain the most silica, and on crystallizing yield free quartz, form a group generally designated the "acid" rocks. Those again that contain least silica and most magnesia and iron, so that quartz is absent while olivine is usually abundant, form the "basic" group. The "intermediate" rocks include those characterized by the general absence of both quartz and olivine. An important subdivision of these contains a very high percentage of alkalis, especially soda, and consequently has minerals such as nepheline and leucite not common in other rocks. It is often separated from the others as the "alkali" or "soda" rocks, and there is a corresponding series of basic rocks. Lastly a small sub-group rich in olivine and without feldspar has been called the "ultrabasic" rocks. They have very low percentages of silica but much iron and magnesia.

Except these last, practically all rocks contain felspars or feldspathoid minerals. In the acid rocks the common feldspars are orthoclase, perthite, microcline, and oligoclase—all having much silica and alkalis. In the basic rocks labradorite, anorthite and bytownite prevail, being rich in lime and poor in silica, potash and soda. Augite is the commonest ferro-magnesian of the basic rocks, but biotite and hornblende are on the whole more frequent in the acid.

Rocks that contain leucite or nepheline, either partly or wholly replacing felspar, are not included in this table. They are essentially of intermediate or of basic character. We might in consequence regard them as varieties of syenite, diorite, gabbro, etc., in which feldspathoid minerals occur, and indeed there are many transitions between syenites of ordinary type and nepheline — or leucite — syenite, and between gabbro or dolerite and theralite or essexite. But, as many minerals develop in these "alkali" rocks that are uncommon elsewhere, it is convenient in a purely formal classification like that outlined here to treat the whole assemblage as a distinct series.

This classification is based essentially on the mineralogical constitution of the igneous rocks. Any chemical distinctions between the different groups, though implied, are relegated to a subordinate position. It is admittedly artificial by it has grown up with the grown of the science and is still adopted as the basis on which more minute subdivisions are erected.

The subdivisions are by no means of equal value. The syenites, for example, and the peridotites, are far less important than the granites, diorites and gabbros. Moreover, the effusive andesites do not always correspond to the plutonic diorites but partly also to the gabbros. As the different kinds of rock, regarded as aggregates of minerals, pass gradually into one another, transitional types are very common and are often so important as to receive special names. The quartz-syenites and nordmarkites may be interposed between granite and syenite, the tonalites and adamellites between granite and diorite, the monzoaites between syenite and diorite, norites and hyperites between diorite and gabbro, and so on.

An atom is composed of a nucleus of protons and neutrons surrounded by an electron cloud. Theoretically, electrons may be found at any distance from the nucleus, although they preferentially rotate around low-energy orbits or *levels*. Within a single level, various *sublevels* can be distinguished. [The term "level" corresponds to "electron shell" in the Bohr model. The terminological analogy is: shell K = level I (n = 1); shell L = level II (n = 2); shell M = level III (n = 3); shell N = level IV (n = 4); and so on.] Electron levels are established according to four quantum numbers:

1. The principal quantum number (n) is related to the energy of the electron levels. An increase in quantum number n corresponds to an increase in the distance of the electron from the nucleus.
2. The orbital or "azimuthal" quantum number (l) defines "form" and indicates which sublevel is occupied by the electron. It assumes integer values between 0 and n–1.
3. The magnetic quantum number (m_l) indicates which orbital of a given sublevel is occupied by the electron. For a given orbital, quantum number l assumes integer values between "1 and +1. An electron with a magnetic quantum number m_l subjected to a magnetic field B increases its energy by a term $E = \mu_b m_l$ B, where μ_b is Bohr's magneton (Zeeman effect).
4. The spin quantum number (m_s) gives the projection of the intrinsic angular momentum of the electron, with values +½ and –½. Two oppositely oriented spins on the same orbital are graphically represented by arrows in opposite directions (↓↑).

The development of quantum mechanics has led to the formulation of various principles governing the structure of atoms:

1. The Aufbau or "build-up" principle states that electrons occupy orbitals of progressively increasing energies.
2. The Pauli Exclusion Principle states that no more than two electrons may occupy the same orbital, and they must have opposite spins. Based on this principle, $2n$ is the maximum number of electrons compatible with a given level.

3. Hund's rule states that, before two opposite-spin electrons can occupy the same orbital of a given sublevel, all the other orbitals of the same sublevel must have been occupied by single electrons of equal spin.

Wavelike Behaviour of Electrons

Like any particle, an electron has a definite mass, corresponding to

$$\frac{1}{1850}$$

of the mass of the hydrogen atom. Nevertheless, experiments with electron beams accelerated by known voltage differences show that a wavelength λ can be associated with the mass of the electron, so that

$$\lambda=\frac{h}{mv}, \quad (1)$$

where h is Planck's constant and v is velocity. Equation 1, which is one of the fundamental results of quantum mechanics, establishes the dual nature of electrons as particles and wavelike elements.

The propagation of a wave in one dimension is obtained by the partial derivative of the wave function

$$\psi = \psi_o \sin 2\pi\left(\frac{x}{\lambda} - vt\right). \quad (2)$$

where ν is frequency (ν = v/λ) and t is time. Function 2 is periodic in t when x is constant (each point vibrates) and periodic in x when t is constant (instantaneous picture of a wave). Partial differentiation of function 2 on x results in

$$\frac{\partial^2\psi}{\partial x^2} = -\frac{4\pi}{\lambda^2}\psi. \quad (3)$$

Principle Quantum Number, *n*	AngularMomentum uantum Number, l	Magnetic Quantum Number, ml	Electron Summary
1	0 (s)	0	1 1s orbital, 2 electrons
2	0 (s)	0	1 2s orbital, 2 electrons
	1 (p)	1, 0, -1	3 2p orbitals, 6 electrons
3	0 (s)	0	1 3s orbitals, 2 electrons
	1 (p)	1, 0, -1	3 3p orbitals, 6 electrons
	2 (d)	2, 1, 0, -1, -2	5 3d orbitals, 10 electrons
4	0 (s)	0	1 4s orbitals, 2 electrons
	1 (p)	1, 0, -1	3 4p orbitals, 6 electrons
	2 (d)	2, 1, 0, -1, -2	5 4d orbitals, 10 electrons
	3 (f)	-3, -2, -1, 0, 1, 2, 3	7 4f orbitals, 14 electrons

The three-dimensional analogue of function 2 is

$$\psi = \psi_0 \sin 2\pi \left[\frac{(x^2 + y^2 + z^2)^{1/2}}{\lambda} - vt \right]. \tag{4}$$

and it can be shown that partial derivation of equation 4 over the three dimensions results in

$$\frac{\partial^2 \psi}{\partial x^2} + \frac{\partial^2 \psi}{\partial y^2} + \frac{\partial^2 \psi}{\partial z^2} + -\frac{4\pi}{\lambda^2}\psi. \tag{5}$$

Let us now consider a particle with mass m, total energy E, and potential energy Φ. The kinetic energy of particle E_k will be

No.	***Element***	***1***	***2***	***3***	***4***	***5***	***6***	***7***
1	Hydrogen	1						
2	Helium	2						
3	Lithium	2	1					
4	Beryllium	2	2					
5	Boron	2	3					
6	Carbon	2	4					
7	Nitrogen	2	5					
8	Oxygen	2	6					
9	Fluorine	2	7					
10	Neon	2	8					
11	Sodium	2	8	1				
12	Magnesium	2	8	2				
13	Aluminium	2	8	3				
14	Silicon	2	8	4				
15	Phosphorus	2	8	5				
16	Sulfur	2	8	6				
17	Chlorine	2	8	7				
18	Argon	2	8	8				
19	Potassium	2	8	8	1			
20	Calcium	2	8	8	2			
21	Scandium	2	8	9	2			
22	Titanium	2	8	10	2			
23	Vanadium	2	8	11	2			
24	Chromium	2	8	13	1			
25	Manganese	2	8	13	2			

Contd...

No.	Element	1	2	3	4	5	6	7
26	Iron	2	8	14	2			
27	Cobalt	2	9	15	1			
28	Nickel	2	8	17	1			
29	Copper	2	8	18	1			
30	Zinc	2	8	18	2			
31	Gallium	2	8	18	3			
32	Germanium	2	8	18	4			
33	Arsenic	2	8	18	5			
34	Selenium	2	8	18	6			
35	Bromine	2	8	18	7			
36	Krypton	2	8	18	8			
37	Rubidium	2	8	18	8	1		
38	Strontium	2	8	18	8	2		
39	Yttrium	2	8	18	9	2		
40	Zirconium	2	8	18	10	2		
41	Niobium	2	8	18	12	1		
42	Molybdenum	2	8	18	13	1		
43	Technetium	2	8	18	13	2		
44	Ruthenium	2	8	18	15	1		
45	Rhodium	2	8	18	16	1		
46	Palladium	2	8	18	18	0		
47	Silver	2	8	18	18	1		
48	Cadmium	2	8	18	18	2		
49	Indium	2	8	18	18	3		
50	Tin	2	8	18	18	4		
51	Antimony	2	8	18	18	5		
52	Tellurium	2	8	18	18	6		
53	Iodine	2	8	18	18	7		
54	Xenon	2	8	18	18	8		
55	Caesium	2	8	18	18	8	1	
56	Barium	2	8	18	18	8	2	
57	Lanthanum	2	8	18	18	9	2	
58	Cerium	2	8	18	19	9	2	
59	Praseodymium	2	8	18	21	8	2	

Contd...

No.	Element	1	2	3	4	5	6	7
60	Neodymium	2	8	18	22	8	2	
61	Promethium	2	8	18	23	8	2	
62	Samarium	2	8	18	24	8	2	
63	Europium	2	8	18	25	8	2	
64	Gadolinium	2	8	18	25	9	2	
65	Terbium	2	8	18	27	8	2	
66	Dysprosium	2	8	18	28	8	2	
67	Holmium	2	8	18	29	8	2	
68	Erbium	2	8	18	30	8	2	
69	Thulium	2	8	18	31	8	2	
70	Ytterbium	2	8	18	32	8	2	
71	Lutetium	2	8	18	32	9	2	
72	Hafnium	2	8	18	32	10	2	
73	Tantalum	2	8	18	32	11	2	
74	Tungsten	2	8	18	32	12	2	
75	Rhenium	2	8	18	32	13	2	
76	Osmium	2	8	18	32	14	2	
77	Iridium	2	8	18	32	15	2	
78	Platinum	2	8	18	32	17	1	
79	Gold	2	8	18	32	18	1	
80	Mercury	2	8	18	32	18	2	
81	Thallium	2	8	18	32	18	3	
82	Lead	2	8	18	32	18	4	
83	Bismuth	2	8	18	32	18	5	
84	Polonium	2	8	18	32	18	6	
85	Astatine	2	8	18	32	18	7	
86	Radon	2	8	18	32	18	8	
87	Francium	2	8	18	32	18	8	1
88	Radium	2	8	18	32	18	8	2
89	Actinium	2	8	18	32	18	9	2
90	Thorium	2	8	18	32	18	10	2
91	Protactinium	2	8	18	32	20	9	2
92	Uranium	2	8	18	32	21	9	2
93	Neptunium	2	8	18	32	22	9	2

Contd...

No.	Element	1	2	3	4	5	6	7
94	Plutonium	2	8	18	32	24	8	2
95	Americium	2	8	18	32	25	8	2
96	Curium	2	8	18	32	25	9	2
97	Berkelium	2	8	18	32	27	8	2
98	Californium	2	8	18	32	28	8	2
99	Einsteinium	2	8	18	32	29	8	2
100	Fermium	2	8	18	32	30	8	2
101	Mendelevium	2	8	18	32	31	8	2
102	Nobelium	2	8	18	32	32	8	2
103	Lawrencium	2	8	18	32	32	8	3
104	Rutherfordium	2	8	18	32	32	10	2
105	Dubnium	2	8	18	32	32	11	2
106	Seaborgium	2	8	18	32	32	12	2
107	Bohrium	2	8	18	32	32	13	2
108	Hassium	2	8	18	32	32	14	2
109	Meitnerium	2	8	18	32	32	15	2
110	Darmstadtium	2	8	18	32	32	16	2
111	Roentgenium	2	8	18	32	32	17	2
112	Copernicium	2	8	18	32	32	18	2
113	Ununtrium	2	8	18	32	32	18	3
114	Ununquadium	2	8	18	32	32	18	4
115	Ununpentium	2	8	18	32	32	18	5
116	Ununhexium	2	8	18	32	32	18	6
117	Ununseptium	2	8	18	32	32	18	7
118	Ununoctium	2	8	18	32	32	18	8

$$E_K = \frac{1}{2}mv^2 = E - \Phi. \tag{6}$$

From equation 6 we obtain

$$m^2v^2 = 2m(E - \Phi). \tag{7}$$

which, when combined with equation 1, gives

$$\frac{1}{\lambda^2} = \frac{2\pi(E - \Phi)}{h^2} \tag{8}$$

Substituting the right side of equation 8 into wave equations 3 and 5 leads, respectively, to

$$\frac{d^2\psi}{dx^2}+\frac{8\pi^2 m}{h^2}(E-\Phi)\psi=0 \tag{9}$$

and

$$\nabla^2\psi+\frac{8\pi^2 m}{h^2}(E-\Phi)\psi=0 \tag{10}$$

where Δ^2 is the Laplacian operator

$$\nabla^2=\frac{\partial^2}{\partial x^2}+\frac{\partial^2}{\partial y^2}+\frac{\partial^2}{\partial z^2} \tag{11}$$

Equation 9 is the *Schrödinger equation for one-dimensional stationary states,* and equation 10 is the tridimensional form. Differential equations 9 and 10 possess finite single-valued solutions, within the domain of squared integrable functions, only for definite values of energy (*eigenvalues*). For vibrational systems, such values are given by

$$E_{(n)}=\left(n+\frac{1}{2}\right)h\nu, \tag{12}$$

where n is the *vibrational quantum number.*

Applying to equation 9 the potential of the *monodimensional harmonic oscillator* (i.e., a particle constrained to a certain equilibrium position by a force that is proportional to its displacement from the origin), we obtain

$$\frac{d^2\psi_{(x)}}{dx^2}+\frac{8\pi^2}{h^2}\left(E-\frac{K_F x^2}{2}\right)\psi_{(x)}=0. \tag{13}$$

If position x of the particle corresponds to its displacement from the origin, the restoring force acting on the particle by virtue of the harmonic potential is given by *Hooke's law:*

$$F=-K_F x, \tag{14}$$

where K_F is the force constant.

Because, in a conservative field, the force is linked to the gradient of the potential by

$$F=-\frac{d\Phi_{(x)}}{dx}, \tag{15}$$

by integration of equation 15 it follows that

$$\Phi_{(x)} = \frac{K_F x^2}{2}, \tag{16}$$

which, substituted into equation 9, leads to equation 13.

The eigenvalues of the energy that satisfy equation 1.13 are given by

$$E_{(n)} = \frac{K_F h^2}{4\pi^2 m}\left(n + \frac{1}{2}\right). \tag{17}$$

We will later adopt function 17 when treating the vibrational energy of crystalline substances and its relationship to entropy and heat capacity.

A more detailed account of the Schrödinger equation can be found in physical chemistry textbooks such as those of Hinshelwood (1951) and Atkins (1978), or in more specialised texts such as that of Hirschfelder et al. (1954). An excellent review of the applications of quantum mechanics to geochemistry has recently been proposed by Tossell and Vaughan (1992).

Periodic Properties of the Elements

The periodic law, discovered by Dmitri Mendeleev, establishes "an harmonic periodicity of properties of chemical individuals, dependent on their masses". To highlight the existence of periodic properties, the various elements are arranged, according to their increasing masses, in groups (8), subgroups (8), and periods (7). A given group comprises elements with similar properties. The periodicity of elemental properties is easily understood if we recall the laws governing the occupancy of atomic orbitals. The periodic table shows the electron structures of the various elements. For the sake of brevity, the structure of the inner shells is indicated by referring to that of the preceding inert gas. An inert gas displays no chemical reactivity (i.e., no tendency to enter into other states of combination). This "stable" configuration corresponds to completion of all sublevels of the principal quantum number *n*. As we can see, a period begins with the occupancy of a new level; elements in this position are ready to lose one electron, with the formation of a univalent positive ion. At the end of the period, immediately preceding the inert gas, the atoms lack a single electron to complete the stable configuration. Such atoms tend to be transformed into univalent negative ions by electron capture.

The number of electrons that complete the period is readily calculated from the following summation:

$$\sum_{l=0}^{n-1}(4l+2)=2n^2 \tag{18}$$

i.e., two electrons in the first period (n = 1), eight electrons in the second (n = 2), 18 electrons in the third (n = 3), and so on.

The first period is completed by only two elements (H, He), because the first level has a single sublevel that can be occupied by two electrons with opposite spin (because $n = 0$, l is also zero; m_l may vary from "l to $+l$ and is also zero; $m_s = +½$ and "½). The second period begins with an electron occupying the lowest energy orbital of sublevel 2s (Li) and terminates with the addition of the sixth electron of sublevel 2p (Ne). The pile-up procedure is analoguous for the third period, but in the fourth period this regularity is perturbed by the formation of transition elements (Sc-Zn) and by progressive pile-up of electrons in the 3d orbitals, whereas the fourth shell is temporarily occupied by one or two electrons. Pileup in the fourth level starts again with Ga, and the period terminates with the completion of the stable octet (Kr).

The same sequence is observed in the fifth period (Y-Cd transition elements). In the sixth period, the transition is observed over sublevel of the fourth shell, generating a group of elements (rare earths, La-Lu) of particular importance in geochemistry for the regularity of progression of reactive properties within the group (the most external shells are practically identical for all elements in the series; the progressive pile-up in the inner fourth shell induces shrinkage of the atomic radius, known as "lanthanidic contraction"). In the seventh period, the transition begins with actinium (Ac) and thorium (Th), whose external electrons occupy sublevel 6 d, and continues with the subsidiary series of protactinium (Pa-Lw), with electrons progressively occupying sublevel of the fifth shell.

Periodic Properties, Ionisation Potentials, and Energy of External Orbitals

In terms of energy, the existence of periodicity is easily understood by analysing the expressions that relate the energy of external orbitals to quantum number n.

Under the *hydrogen ion approximation* (ions of nuclear charge Ze with a single electron progressively occupying the various orbital levels), the potential acting on the electron by virtue of the nuclear charge is given by

$$\Phi_C = -\frac{Ze^2}{r}, \tag{19}$$

where Z is the atomic number and r is the distance from the nucleus expressed in Cartesian coordinates:

$$r = \left(x^2 + y^2 + z^2\right)^{1/2}. \tag{20}$$

Substitution of equation 1.19 into equation 1.9 leads to

$$\nabla^2\psi + \frac{8\pi^2 m}{h^2}\left(E + \frac{Ze^2}{r}\right)\psi = 0. \tag{21}$$

The eigenvalues of equation 1.21 are

$$E_{(n)} = -\frac{2\pi^2 me^4 Z^2}{n^2 h^2}, \tag{22}$$

where $E_{(n)}$ is the energy of the various orbital levels in the hydrogen ion approximation.

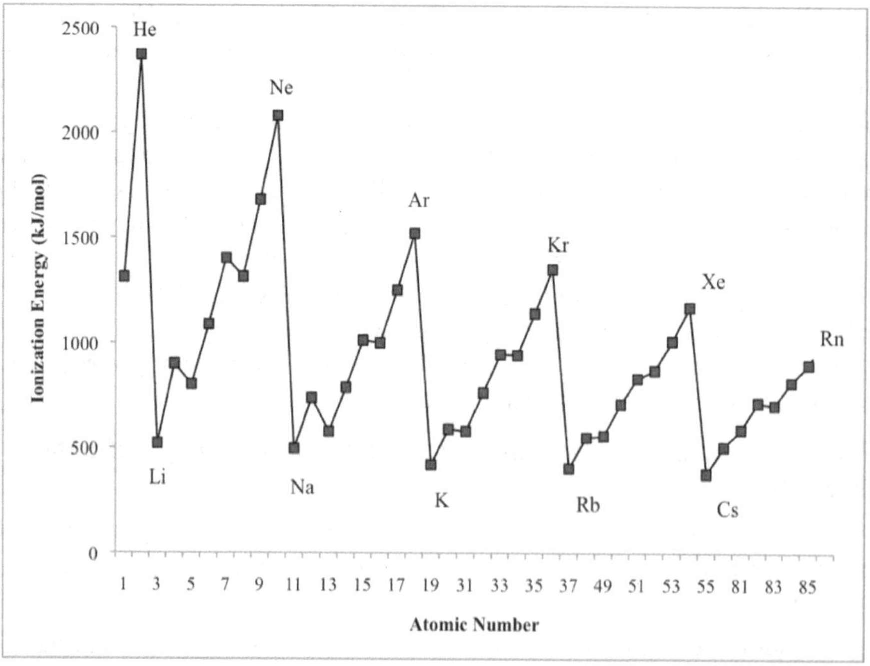

Figure: *First Ionisation potentials as functions of atomic number.*

The energy of external orbitals in real atoms (which is numerically equivalent to the first Ionisation potential expressed in eff) is reproduced by an empirical expression of the form:

$$E = \frac{(Z - S_e)^2 C_e}{n^2} = \frac{Z_{eff}^2 C_e}{n^2} \tag{23}$$

where S_e is a screen constant, depending on the number of inner electrons, and C_e is a conversion factor. Equation 23 establishes that, with an increase in the quantum number, we should observe a jump in the energy values of the external orbitals. We recall for this purpose that the external or "valence" electrons are those involved in the formation of atomic bonds (because the absolute values of their orbital energies are much lower than those pertaining to the inner electrons). The first Ionisation potentials plotted as functions of atomic number Z exhibit marked discontinuities, corresponding to the variations in the principal quantum number, and hence corresponding to the change of period in Mendeleev's periodic chart. Within the same period, we ob serve a progressive increase in the first Ionisation potential, with minor discontinuities for transition elements (due to the progressive filling of inner shells, which modifies the value of screen constant S_e in eq. 23). Recalling the definition of "valence electron" and considering that bonding energies are related to external orbitals, the periodicity of the reactive properties of the elements is readily understood in light of equations 22 and 23.

Generalities on the Concept of Chemical Bond in Solids

"There is a chemical bond between two atoms or groups of atoms in the case that the forces acting between them are such as to lead to the formation of an aggregate with sufficient stability to make it convenient for the chemist to consider it as an independent molecular species" (Pauling, 1948). We will consider four distinct types of bonds:

1. ionic bonds
2. covalent bonds
3. van der Waals bonds
4. metallic bonds

Although in nature it is rare to see solid compounds whose bonds are completely identifiable with one of the types listed above, it is customary to classify solids according to the dominant type of bond—i.e., *ionic, covalent, van der Waals,* and *metallic* solids.

In ionic solids, the reticular positions are occupied by ions (or anionic groups) of opposite charge. In NaCl, for instance, the Cl^- ions are surrounded by six Na^+ ions. The alternation of net charges gives rise to strong coulombic interactions.

In covalent solids, two bonded atoms share a pair of electrons. Covalent bonds are particularly stable, because the sharing of two electrons obeys the double duty of completing a stable electronic configuration for the two neighbouring atoms. In diamond, for instance, each carbon atom shares an electron pair with four surrounding carbon atoms, generating a stable giant molecule that has the dimension of the crystal.

In van der Waals solids, the reticular positions are occupied by inert atoms held together by van der Waals forces. The van der Waals bond is rather weak, and the bond does not survive a high thermal energy (*Fm3m* solid Ne, for instance, is stable below 20 K). The van der Waals bond also exists, however, in molecular compounds (e.g., solid CO_2). The reticular positions in this case are occupied by neutral molecules (CO_2). Within each molecule, the bond is mainly covalent (and interatomic distances are considerably shorter), but the various molecules are held together by van der Waals forces whose nature will be discussed later.

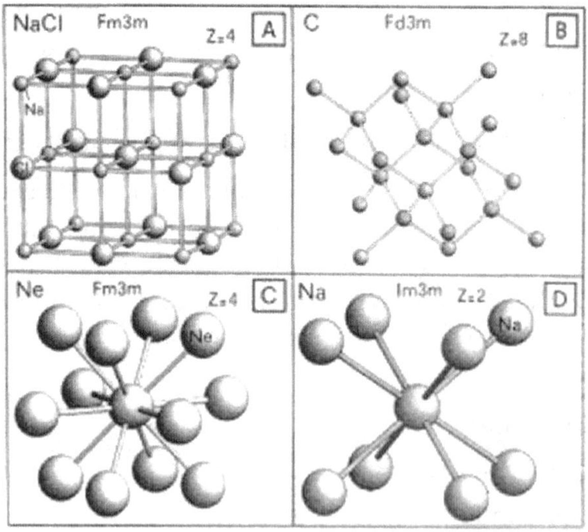

Figure: *Various types of solids. For each solid, a sketch of the bond is given.*

In metallic solids, the reticular positions are occupied by cations immersed in a cloud of delocalised valence electrons. In solid Na, for instance, the electron cloud or "electron gas" is composed of electrons of the *s* sublevel of the third shell. Note that the type of bond does not limit the crystal structure of the solid within particular systems. For example, all solids belong to the cubic system, and two of them (NaCl, Ne) belong to the same structural class (*Fm3m*).

Ionic Bond and Ionic Radius

The concept of "ionic radius" was derived from the observation of crystal structures understood as "regular arrays of ions of spherical undeformable symmetry." The structure of ionic crystals obeys pure geometrical rules and corresponds to the energetically optimal condition of an ion surrounded by the maximum number of ions of opposite charge.

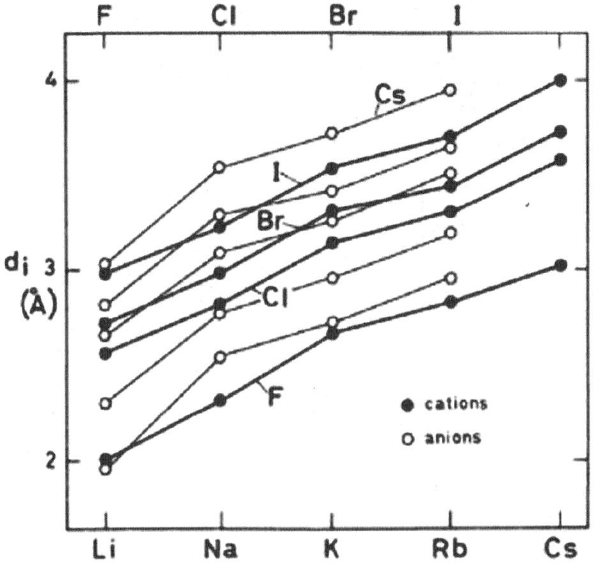

Figure: *Interatomic distances in alkali halides, arranged in cationic and anionic series.*

The assumption of "undeformability" makes it possible to consider interionic distance as a constant, at parity of ionic charge. The concept of ionic radius is thus related to that of "interionic distance" (d_i) by simple addition, according to

$$d_i = r_+ + r_-, \tag{24}$$

where r_+ and r_- are, respectively, the ionic radii of a cation and an anion at the interionic distance observed in a purely ionic crystal.

The concept that the ionic radius is relatively independent of the structure of the solid arose intuitively from experimental observations carried out on alkali halides, which are ionic solids par excellence.

Significant parallelism within each of the two families of curves may be noted. This parallelism intuitively generates the concept of constancy of the ionic radius.

The first attempt to derive empirical ionic radii by geometrical rules applied to ionic compounds (and metals) was performed by Bragg (1920), who proposed a set of ionic radii with a mean internal precision of 0.006 Å. Further tabulations were proposed by Lande (1920), Wasastjerna (1923), and Goldschmidt et al. (1926).

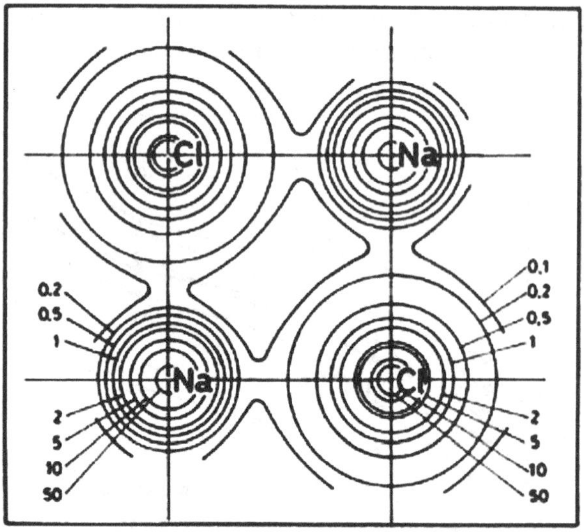

Figure: *Electron density (in electrons per Å^3) in the (100) plane of NaCl.*

However, we have already seen that the Schrödinger equation denies the existence of definite limits in the geometry of atomic orbitals, and that the presence of an electron in a particular portion of the space surrounding the nucleus is a probabilistic concept. These two facts apparently conflict with the concept of an ion as a spherical undeformable entity. But this conflict is only apparent, as may be deduced from the "electron density maps" obtained for crystalline compounds by X-ray diffraction techniques. Concentric lines delineate zones of equal electron density (electrons per Å^3), increasing toward the nuclei (to a maximum of 70 electrons per Å^3 for Na^+ and 130 electrons per Å^3 for Cl^-). The absolute minimum electron density between the two nuclei defines the limits (hence the "dimensions") of the two ions. By integration of the electron densities, it can be calculated that 17.70 electrons are present in the region surrounding the nucleus of Cl^- and 10.05 electrons are present in the region surrounding the nucleus of Na^+.

These values, compared with the theoretical limits of 18 and 10 electrons, respectively, indicate that, within experimental approximation, there has been the complete transfer of one electron—and, hence, that a completely "ionic bond" is present in the cubic structure of NaCl. We also observe that the electron density map defines almost perfect sphericity for both ions. The "ionic radii" of Na^+ and Cl^- in the crystalline cubic compound NaCl thus correspond to the radii (radial extensions) of the two elements in the condition of "ionic bond" when there is complete transfer of one electron from Na to Cl.

Pauling's Univalent Ionic Radii

The first attempt to derive ionic radii independently of the geometrical concept of simple additivity was made by Pauling (1927a). His method was based on the assumption that the dimension of an ion is determined by the distribution of its outer electrons. As we have already seen (eq. 23), the "effective nuclear charge" of an ion (Z_{eff})is lower than its atomic number, due to the shielding effect of inner electrons, expressed by screening constant S_e. Values of S_e may be obtained theoretically, and partially by results of X-ray diffraction studies. Based on such values, the univalent ionic radius (r_{+-}) can be expressed as

$$r_{+-} = \frac{C_n}{Z - S_e} = \frac{C_n}{Z_{eff}}, \tag{25}$$

where C_n is a constant determined by the principal quantum number of the outermost electrons in the inert gas configuration. Applying equation 25 to couples of isoelectronic ions such as NaF, KCl, RbBr, and CsI, which have, respectively, the same C_n, Pauling determined the dimension ratios of the isoelectronic ions in the corresponding solids and derived the corresponding ionic radii by combining obtained with observed values of interionic distance. For ions of the neon structure, for example, S_e has a value of 4.52, and the values of Z_{eff} for Na^+ and F^- are thus 6.48*e* and 4.48*e*, respectively. Because C_n is identical for the two ions, knowledge of the interionic Na^+–F^- distance (2.31 Å) gives the univalent radii solving the system

$$\frac{r_{F^-}}{r_{Na^+}} = \frac{Z_{eff,Na^+}}{Z_{eff,F^-}} = \frac{6.48}{4.48}, \tag{26}$$

$$r_{Na^+} + r_{F^-} = 2.31 \tag{27}$$

which gives 36 Å for F⁻ and 0.95 Å for Na^+. If we now continue to apply equation 1.25 to multivalent ions, assuming that C_n remains unchanged at the value of the intermediate inert gas, we obtain values for all ions with He, Ne, Ar, Kr, and Xe structures. These radii, known as "univalent ionic radii in VI-fold co-ordination," "do not have absolute values such that their sums are equal to equilibrium inter ionic distances" but rather "are the radii the multivalent ions would possess if they were to retain their electron distributions but enter into Coulomb interaction as if they were univalent". Univalent radii may be opportunely modified to obtain effective ionic radii in real structures:

$$r_Z = r_{+-} Z^{-2/(n-1)}, \qquad (28)$$

where n is the Born exponent and z is the ionic charge.

Covalent Bond and Covalent Radius

According to Lewis's (1916) definition, a "covalent bond" exists whenever two atoms are linked in a stable molecule by sharing two outer electrons. We distinguish between "homopolar" and "heteropolar" covalent bonds on the basis of whether or not the bonded atoms are of the same type. Pauling (1960) has shown that in covalent compounds the distance between two elements A and B, independent of the nature of the two atoms, is the same in all compounds containing both elements, regardless of the state of aggregation of the compound (i.e., solid, liquid, or gaseous). Covalent radii are easily calculated by dividing by 2 the interatomic distances observed in homopolar covalent compounds (e.g., P_4, S_8, etc.). An internally consistent tabulation was derived by Sanderson (1960) from interatomic distances in heteropolar covalent gaseous molecules (e.g., $AlCl_3$, CH_4, SiF_4, etc.). The covalent radii of Sanderson are defined as "nonpolar," because they are obtained from compounds that do not have charge polarity.

Electronegativity and Fractional Ionic Character

The concept of "electronegativity" is widely used in geochemistry to define the combinational properties of the various elements in a qualitative manner. However, it is important to stress that this concept is largely empirical, because it varies from author to author and cannot be univocally quantified.

According to Pauling's (1932) definition, electronegativity is "the capacity of an atom of a molecule to attract the electrons of another atom." The electronegativity scale of Pauling (1960) is based on the differences in the binding energies in diatomic molecules AB with respect to homeopolar compounds AA and BB, according to

$$Z_A - Z_B = \gamma^{1/2}\left[E_{AB} - \frac{1}{2}(E_{AA} + E_{BB})\right]. \quad (29)$$

where χ_A and χ_B are the electronegativities of atoms A and B, E_{AB}, E_{AA}, and E_{BB} are the binding energies of compounds AB, AA, and BB, and operative constant γ is a term that constrains the adimensional χ_A and χ_B to change by 0.5 with each unit valence change in the first row of the periodic table ($\gamma^{1/2}$ = 0.208).

According to Mulliken (1934, 1935), electronegativity is the algebraic mean of the first ionisation potential and of electron affinity.

According to Gordy (1946), electronegativity is represented by the value of the potential resulting from the effect of the nuclear charge of an unshielded atom on a valence electron located at a distance corresponding to the covalent radius of the atom.

According to Sanderson (1960), electronegativity represents the mean electron density per unit volume of an inert element:

$$D_C = \frac{3Z}{4\pi_C^3}, \quad (30)$$

where Z is the atomic number and r_c is the covalent nonpolar radius.

Iczkowski and Margrave (1961) consider electronegativity as the local value of a potential defined by the partial derivative of the energy of the atom W with respect to ionic charge z:

$$Z_{(z)} = \frac{\partial W}{\partial z}. \quad (31)$$

The most immediate use of the concept of electronegativity is in the definition of "fractional ionic character" ($|_i$). The more or less ionic character of a bond is determined by the distribution of charges within the diatomic molecule. For a purely ionic bond, there is complete charge transfer between the two neighbouring atoms, whereas in a purely covalent bond there is no charge transfer. The relationship between electronegativity and percentage of ionicity in diatomic molecules has been proposed by several authors in different ways. According to Pauling (1932, 1960), the more or less ionic character of a bond is expressed by parametre $|_i$, derived as follows:

$$f_i = 1 - \exp\left[-\frac{1}{4}(Z_a - Z_B)^2\right]. \quad (32)$$

According to Gordy (1950), the same parametre can be obtained by applying

$$f_i = \frac{1}{2}(Z_A - Z_B). \tag{33}$$

Hinze et al. (1963) suggest that charge transfer in diatomic molecules obeys the principle of equality of energies involved in the process. If we adopt equation 31 as a definition of electronegativity and imagine that the amount of energy involved in the transfer of a fraction of charge *dz* from atom A to atom B is quantifiable as

$$dE_A = \left(\frac{\partial W_A}{\partial z}\right)_{ZA}^{dZ} \tag{34}$$

then, by application of the equality principle, we have

$$dE_A = \left(\frac{\partial W_A}{\partial z}\right)_{ZA} dz = \left(\frac{\partial W_B}{\partial z}\right)_{ZB} dz = dE_B. \tag{35}$$

A similar concept was developed by Sanderson (1960).

Electronegativity values according to Pauling and Sanderson are listed together with the first four Ionisation potentials and the electron affinities of the various elements.

At present, the most widely used definition of fractional ionic character in solid state chemistry is that of Phillips (1970), based on a spectroscopic approach. Phillips defined fractional ionic character as

$$f_i = \frac{E_i^2}{E_i^2 + E_c^2}, \tag{36}$$

where E_i ("ionic energy gap") is obtained from the "total energy gap" (E_g) between bonding and antibonding orbitals:

$$E_g = \frac{\hbar\omega_r}{(\varepsilon_{\_} - 1)^{1/2}} \tag{37}$$

$$E_i = \left(E_g^2 - E_c^2\right)^{1/2} \tag{38}$$

where $\hbar\omega_p$ is the plasma frequency for valence electrons, ε_∞ is the optic dielectric constant, and E_c corresponds to E_g for the nonpolar system in the same row of the periodic table, with a correction for interatomic spacing.

Catlow and Stoneham (1983) have shown that ionic term E_i corresponds to the difference of the diagonal matrix elements of the Hamiltonian of an AB molecule in a simple LCAO approximation

$(H_{AA} - H_{BB})$, whereas the covalent energy gap corresponds to the double of the off-diagonal term H_{AB}—i.e.,

$$E_g = \sqrt{(H_{AA} - H_{BB})^2 + 4H_{AB}}. \tag{39}$$

Ionic Polarizability and van der Waals Bond

The electron cloud of an ion subjected to an electric field undergoes deformations that may be translated into displacement "*d*"of the baricentres of negative charges from the positions held in the absence of external perturbation, which are normally coincident with the centres of nuclear charges (positive).

The noncoincidence of the two centres causes a "dipole moment," determined by the product of the displaced charge (Z_e) and the displacement *d*.

The displacement is also proportional to the intensity of the electrical field (F). The proportionality factor (α) is known as "ionic polarizability":

$$\mu_d = Z_e d = F\alpha. \tag{40}$$

Let us consider two neighbouring atoms A and B and imagine that the electronic charge distribution is, for an infinitesimal period of time, asymmetrical. In this period of time, atom A becomes a dipole and induces a distortion in the electron cloud of atom B by attracting electrons from its positive edge: atom B is then transformed, in turn, into a dipole. All the neighbouring atoms undergo the same process, generating attractive forces among themselves.

Due to electron orbital motion, the induced forces fluctuate with a periodicity dictated by the rotational velocity of electrons. The *van der Waals bond* is the result of the induced attractions between fluctuating dipoles.

The attractions diminish with increased molecular distance and, when distances are equal, increase with the number of electrons within the molecule.

Free-ion polarizability calculations have been proposed by various authors, based on the values of molecular refraction of incident light (R_m), according to

$$\alpha = \frac{3R_m}{4\pi N_0}, \tag{41}$$

where N_0 is Avogadro's number.

Free-ion polarizabilities, arranged in isoelectronic series, are listed. A semilogarithmic plot of the listed values reveals a functional dependence on atomic number. This dependence, quite marked, allows us to estimate the free-ion polarizability for ions for which there are no precise experimental data.

Crystal Radius and Effective Distribution Radius (EDR)

The first satisfactory definition of crystal radius was given by Tosi (1964): "In an ideal ionic crystal where every valence electron is supposed to remain localised on its parent ion, to each ion it can be associated a limit at which the wave function vanishes. The radial extension of the ion along the connection with its first neighbour can be considered as a measure of its dimension in the crystal (crystal radius)."

This concept is clearly displayed in which the radial electron density distribution curves are shown for Na^{+} and Cl^{-} ions in NaCl. The nucleus of Cl^{-} is located at the origin on the abscissa axis and the nucleus of Na^{+} is positioned at the interionic distance experimentally observed for neighbouring ions in NaCl.

The superimposed radial density functions define an electron density minimum that limits the dimensions or "crystal radii" of the two ions. We also note that the radial distribution functions for the two ions in the crystal (continuous lines) are not identical to the radial distribution functions for the free ions (dashed lines).

The "crystal radius" thus has local validity in reference to a given crystal structure. This fact gives rise to a certain amount of confusion in current nomenclature, and what it is commonly referred to as "crystal radius" in the various tabulations is in fact a mean value, independent of the type of structure.

The "crystal radius" in the sense of Tosi (1964) is commonly defined as "effective distribution radius" (EDR). The example given shows radial electron density distribution curves for Mg, Ni, Co, Fe, and Mn on the M1 site in olivine (orthorhombic orthosilicate) and the corresponding EDR radii located by Fujino et al. (1981) on the electron density minima. It also must be stressed that EDR may assume different values along various crystallographic directions. This is clearly evident if we compare electron density maps obtained by calculation of electron density functions of neutral, spherically averaged atoms (known as Independent Atom Model or IAM) with observed total electron densities.

For instance, makes this comparison for BeO: subtraction of IAM densities (B) from total electron densities (A) results in a deformation density chart (C), which shows how the electron density of the IAM reference model has migrated due to bonding effects. (Solid contours in the figure are positive accumulations of charges; short dashes are negative; long dashes mark unperturbed zones.)

2

Geoscience Conceptual Knowledge

Most instructors recognise that students enter their classes with at least some prior knowledge, understanding, and beliefs. Effective learning requires students to integrate new information with this prior knowledge, and in many cases, to correct prior knowledge that is incomplete or incorrect. Effective learning also requires students to move from what may be rote, disconnected knowledge to a deeper, interconnected understanding of key concepts. An important goal of effective teaching, therefore, is to ensure that students progress from incomplete or incorrect ideas to a deeper and more correct conceptual understanding of the topic. Consequently, instructors should be aware of student prior knowledge to ensure effective teaching. Although many instructors recognise the role and importance or prior knowledge in learning, it is not always easy to ascertain what students know. Moreover, instructors need to know if student ideas have changed in response to teaching. Clearly, if the goal of instruction is to have students leave a course with greater and more correct conceptual understanding, tools are needed to assess this.

Libarkin and Kurdziel (2001) reviewed literature related to student prior knowledge and alternate conceptions and made several recommendations for how college instructors can assess student ideas. Open-ended, pre-instruction questionnaires and concept maps can be highly effective tools for assessing student ideas, yet these open-ended tools can be time-consuming for instructors to interpret and use. Multiple-choice tests in which the distracters (incorrect responses) are drawn from research on student alternate conceptions have the potential to provide instructors with more rapid yet still useful feedback as to what pre-instruction ideas their students hold. Other science disciplines have taken this approach and have seen the development and use of such multiple-choice tools for the purposes of assessing student knowledge prior to instruction, measuring student learning gains in a particular course, and determining the effectiveness of different instructional approaches. In physics, the Force Concept Inventory is widely used to assess student learning in introductory courses. Other disciplines have followed suit, with research-based concept inventories now available in chemistry, astronomy, and biology.

The need for a tool to assess student knowledge of concepts commonly taught in introductory, college-level geoscience courses has recently been addressed by Libarkin and Anderson through development of the Geoscience Concept Inventory (GCI). Test questions (items) on this multiple-choice instrument were developed from extensive analysis of actual student responses to open-ended questions covering a range of geoscience content, and validated using statistical tests. Libarkin and Anderson (2005) demonstrated the utility of the GCI in assessing student learning and the effectiveness of instruction across a variety of college types, locations, and specific introductory courses. Elkins and Elkins (2007) also used the GCI to assess student learning and instruction in an entirely field-based course. In this study, we utilise the GCI to examine conceptual knowledge of geoscience topics among preservice elementary teachers. Specifically, we use the GCI to (1) examine the level of prior geoscience conceptual knowledge of future elementary teachers, and (2) assess changes in conceptual knowledge among future elementary teachers during instruction in an introductory geoscience course specifically designed for this population. Whereas the goal of teaching is to increase learning among all students, it is especially important that future teachers gain a deep understanding of geoscience content so that they can effectively teach their future students.

Student Conceptions in the Geosciences

Student learning and conceptual knowledge in the sciences is an active field of research. Although theorists disagree on how exactly learning takes place, authors of most of this literature agree on several main points: (1) the knowledge, skills, experiences, and beliefs that students bring into the classroom exert a powerful effect over what students learn and retain from their classroom experiences; (2) students' prior ideas (variously termed "misconceptions," "prior conceptions," or "alternate conceptions") are deeply entrenched and resistant to change; and (3) appropriately structured learning experiences that take into account student prior knowledge can lead to meaningful learning. In addition to studies of how prior knowledge interacts with learning, some of the research in this held focuses on cataloguing student ideas in particular fields, including earth science. The extensive alternative conceptions literature is a useful starting point for instructors who want to understand what ideas their students may hold, and provides a valuable framework from which instructors can build meaningful instruction.

Studies of student conceptions in the geosciences have largely focused on pre-college populations. Recent work has begun to examine geoscience conceptions among college students, as well as preservice and inservice teacners (populations most similar to ours). However, as noted by Orion and Ault (2007), many of the same preconceptions appear across grade levels K-16 and among adults, and many K-12 teachers hold the same non-scientific conceptions as their students. This research has revealed an array of student prior knowledge and conceptions, and has identified specific area of conceptual difficulty that can be organised around themes common to introductory geoscience courses.

Given the diversity of prior conceptions (and conceptions retained by students despite instruction) reported in the literature, it would be exceedingly difficult for a college instructor to ascertain and address each individual idea held by each student in his or her class. Research-based multiple-choice instruments such as the GCI, therefore, can provide rapid but valuable feedback as to what conceptions students hold and how (or even if) these ideas change during instruction

Classroom Context

All study participants were enrolled in an entry-level physical and historical geology course at a mid-sized, Midwestern university. The course, Earth Science for Elementary Educators II (ESEE2), is specifically designed for undergraduate students pursuing teacher certification programs in elementary education or in integrated elementary science (certification for teaching grades K-8, children aged 5-14). This course is one of six entry-level science courses in the subject areas of life, physical and earth science. It is also one of two earth science courses offered in the program; the other is an entry-level physical geography course (Earth Science for Elementary Educators I; ESEE1). Students pursuing elementary teaching certification must take four of the six science courses, with at least one course from each subject area. Students pursuing the elementary integrated science certification must take all six science courses.

The course format consists of two, 2 hour and 20 minute sessions per week. In-class student learning occurs predominantly through laboratory activities, small group discussions and written assignments, and brief lectures integrated with whole class discussion. Regular readings and homework assignments accessed via an on-line system supplement the in-class learning and provide opportunities for additional instructor feedback. Enrollment is limited to 24 students per course section; typically 2 course sections using identical curricula are offered per semester. Sections are taught by a geoscience/science education faculty member, geoscience education doctoral students, and a part-time instructor. Although this is a content course, instructors utilise methods such as inquiry teaching to model teaching practices appropriate to an elementary classroom. In addition to geoscience content, the course explicitly addresses topics of interest to future elementary teachers, such as children's ideas and misconceptions in earth science, state and national earth science benchmarks, and lesson plans and activities appropriate for K-8 students..

Instrumentation

The GCI - Development and evaluation of the GCI is detailed in Libarkin and Anderson (2006, 2008) and summarised here. The GCI was developed through an iterative process of qualitative data collection, item development and expert review, pilot testing, statistical analysis, item revision, and further testing. Although the topical content of the GCI was predetermined by the authors, individual items were derived from open-ended surveys and interviews of more than 1000 students from a wide variety of college settings. Utilising a grounded theory methodology, the authors mined interview and survey data for fundamental concepts and authentic student responses, thus ensuring that items were based on existing student conceptions (as well as misconceptions), rather than rote knowledge.

Items were then administered as a pretest to more than 3500 students, reviewed by both content and education external experts, and reviewed by faculty participating in the pilot testing. Testing and feedback guided revision of individual items; revised items were then administered to participating students as a post-test. Follow-up think-aloud interviews, in which students were asked to explain their responses to individual items, were then used to gauge student interpretation of questions and the extent to which questions evaluated alternate conceptions. Together, this iterative development procedure established the construct and content (face) validity of the instrument. Additional validity and reliability measures, as well as the scaling function for the GCI, were determined by Rasch and differential item functioning (DIF) analysis. From a theoretical perspective, the Rasch approach, a type of Item Response Theory (IRT) analysis, assumes that: (1) test item statistics will fit the Rasch model, (2) student performance on the test instrument will be improved once a concept is learned, (3) the characteristics of specific test items are independent of the test subject's ability, and (4) that both overall test and individual item performance are directly related to the trait being measured (conceptual understanding of the geosciences).

The Rasch analysis also considers the difficulty of individual items on the test, and the scaling function developed from the statistical analysis correlates test score directly with a subject's conceptual understanding. Thus a subject's scaled test score and response to particular items directly correlates to his or her level of conceptual understanding of these topics.

These development and validation procedures have resulted in the current version of the GCI, which includes 73 items arranged into 11 difficulty categories. Users of the GCI construct a subtest by choosing a question from each of the 11 categories plus 4 questions required of all subtests, resulting m a 15-item instrument. The GCI is developed such that although an individual user can tailor items to the content of his or her courses, the scoring of the instrument is tied to a single scale. This allows users to compare student learning and instructional approaches nationwide.

The GCI subtest used in this study was constructed from the 4 required anchor items (1, 2, 37, and 73) plus 1 item selected from each of the 11 difficulty categories. Items chosen were representative of content addressed in the course, with the majority of the items directly related to content encountered by students in the laboratory activities or class discussions. Thus the test items we chose represent what we believe are some of the key concepts covered by the course.

Research Procedure

The 15-item GCI was administered as a pretest in each participating course section during the first week of the semester. An identical version of the GCI was administered 15 weeks later during the last week of the course. In general, students took 15-20 minutes to complete the test. The course instructors did not view participant results until after each semester had ended.

The GCI was administered to a total of six course sections of ESEE2 over three semesters. One author (Petcovic) taught five of the course sections, and a part-time instructor taught the remaining section. Due to a realignment of course objectives between ESEE1 and ESEE2, significant content revisions took place between Spring 2005 and Fall 2005. These included the introduction of several new activities particularly in the area of plate tectonics, introduction of the online homework component, and introduction of a semester project where students investigate children's misconceptions of specific earth science topics. Because the content of the course changed, we replaced three GCI items to better reflect revised course content. The new items were of equivalent difficulty to the removed items, thus preserving the overall scoring of the GCI. Minor revisions to the homework and semester project components occurred between Fall 2005 and Spring 2006, but the GCI test was not change during this time.

Study Population

ESEE2 students present on the testing days and consenting to participate in the study took the GCI. Based on demographic data collected during the pretest, the aggregate study population (n = 122) is about 85% female, 98% white, traditionally-aged college juniors and seniors (mean age 21.8). Individual course sections range from 74-90% female. Approximately three-quarters of participants reported that they had taken an earth science course in high school, and 95% reported that they had taken another earth science course in college. Nearly all of these students have taken ESEE1 or the equivalent general education course in physical geography covering topics of atmosphere, weather, and climate. Only 9 students reported having taken a prior college-level course in physical or historical geology.

Overall Results of the GCI

The mean scaled score on the GCI pretest among all participants (n = 122) was 43.9%, with individual scores ranging from 20.3% to 67.0% correct on both the pretest and posttest. However, even though the range or scores was similar between the pre- and posttests, the mean scaled score on the posttest (n = 102) was 48.1%, showing an overall gain (pre to post) among all participants of 4.2%. Students in sections 3-6 had a slightly greater overall mean gain of 6.5%. Improvements were seen among both high- and low-scoring students, but the greatest normalised gains were seen among students who scored 30-40% correct on the pretest. We were able to match pretests and posttest for 89 individual students; thus subject attrition between the pretest and posttest was 27%. Attrition was due to either students being absent on a testing day, or students choosing not to consent to participate in the study.

Student responses to individual GCI items ranged from 0-83% correct on the pretest, and 0-92% correct on the posttest. Items covered a wide range of difficulty, as determined by fraction of correct responses on the pretest; 50% or more of pretest responses were correct for about half of the test items, and fewer than 50% of the pretest responses were correct for the remaining test items. Pre- to posttest gains occur among both easy and difficult test items.

We analysed the GCI data collected during this study quantitatively to determine (1) whether statistically significant differences on pretest performance existed between ESEE2 course sections, (2) whether observed pretest to posttest changes in mean GCI scores were statistically significant for individual course sections, and (3) for course sections with significant pretest to posttest change, which individual GCI items were causing this change. All analyses utilised matched pretests and posttests for individual students (n = 89).

One-way Analysis of Variance (ANOVA) tests on pretest scores were used to determine whether significant differences in pretest performance, and thus the initial level of geoscience conceptual knowledge, existed between course sections. A one-way repeated measures ANOVA on pretest to posttest difference scores using course section as a covariate was performed to determine whether course section had a significant effect on student performance. Significant results prompted further testing, so paired sample t-tests on the difference scores for each course section were performed to determine whether pre- to posttest performance changes were significant within any of the course sections. Thus, these tests track overall pretest to posttest learning as measured by the GCI. For these analyses, a p-value of 0.05 or less (95% or greater level of confidence) indicated significance.

For the course sections in which the previous tests revealed a statistically significant overall change in mean pretest to posttest difference score, we analysed the fraction of students responding correctly to each GCI items using the McNemar chi-square test for matched pairs. The chi-square test enabled us to identify individual

GCI items for which we observe a statistically significant change between the pretest and posttest, hence indicating a significant change in students' level of knowledge of individual GCI topics.

Analysis of Pretest GCI Scores

Reports the mean pretest, posttest, and matched pre-to posttest difference scores for all six course sections. The p-value obtained from the one-way ANOVA performed on all raw pretest scores was 0.54, indicating that there were no statistically significant differences on the pretest among the course sections. An effect size of $7^2 = 0.045$ was observed, thus the section factor explains 4.5% of the variance in the pre-test scores. The initial level of geoscience concept knowledge as measured by the GCI was therefore similar across all six course sections.

Analysis of Pretest to Posttest Mean GCI Difference Scores

The repeated measures ANOVA results revealed a lack of significant difference between pretest and posttest scores for matched pairs of tests, as indicated by an insignificant p-value for test effect of 0.364, with an effect size of; $7^2 = 0.009$. However, the section covariate had a significant interaction with the difference scores, as verified by the p-value of 0.006 with an effect size of $7^2 = 0.084$. The average posttest score was higher than the average pretest score for five of the course sections. Further analysis of paired sample t-tests revealed that the pre- to posttest difference in mean scores was significant for three of the course sections (3, 4, and 5). The only course section for which the average posttest score was lower than the average pretest score was section 1; however, the paired sample t-test indicated that this drop was not statistically significant.

Analysis of Correct Response to Each GCI Item

McNemar chi-square tests for matched pairs were performed for all 15 test items for the three course sections in which we found statistically significant pretest to posttest improvement. Based upon the chi-square test results, statistically significant increases in the number of correct responses were observed for specific items in course sections 3 (GCI items 28, 70 and 73), 4 (items 21 and 28), and 5 (item 14). Thus, it appears that these items were responsible for the significant increase in overall posttest performance within these sections. Interestingly, we observed a statistically significant decrease in the number of correct responses to item 2 in Section 5, although this section did have an overall mean pretest to posttest increase.

Our study has utilised the GCI to (1) examine the level of prior geoscience conceptual knowledge of future elementary teachers, and (2) assess pretest to posttest changes in conceptual knowledge among future elementary teachers enrolled in ESEE2. Thus our analysis of GCI results serves as a tool to assess both the prior knowledge of our students, and the efficacy of our course designed for this population of students. We acknowledge that using a multiple-choice instrument to examine student ideas is limited; however well-designed, a multiple-choice test is unlikely to reveal the full extent to which students truly understand concepts, and the extent to which these concepts are interconnected to create conceptual frameworks. However, because of the way in which it was developed and validated, the GCI is particularly useful in assessing whether students have correct conceptual (as opposed to rote) knowledge of particular geoscience topics.

Prior Geoscience Knowledge of Preservice Elementary Teachers

Overall, our students performed poorly on the GCI pretest, with a mean scaled score of 43.9%. Other users of the GCI report similar scaled pretest scores of 39-43%, suggesting that our population of preservice elementary teachers have a pre-instructional level of geoscience knowledge that is similar to the national sample of undergraduate students, as measured by the GCI. Additionally, we did not find any statistically significant differences on mean pretest scores between course sections, suggesting that our student population as a whole has similar levels of pre-college geoscience preparation.

A widespread perception exists that future elementary teachers have weak academic backgrounds in science. Research comparing the academic science preparation of preservice elementary and secondary teachers and comparing earth science content knowledge of preservice and inservice teachers does confirm that preservice elementary teachers typically are weaker in science than other populations of students. However, the results of our study suggest that our population of preservice elementary teachers has pre-instructional knowledge of geoscience equal to the national sample of undergraduates reported by Libarkin and Anderson (2005).

Until 2006, earth science was a high school graduation requirement in Michigan, where most of our students attended high school. We suggest that the similarity in pretest performance between our students and the national sample of undergraduates reflects that most of our students (about 75%) have taken an earth science course in high school. Our student population of future elementary teachers, clearly, is no weaker in earth science than the national sample of undergraduates, based on GCI pretest score.

Overall Changes in Geoscience Knowledge among ESEE2 Students

The overall gains made by students enrolled in our course are similar to the average scaled gain of 4% reported by Libarkin and Anderson (2005), which utilised a 20-question version of the GCI to examine over 2500 students enrolled in 29 courses nationwide. Among the significant findings from their study, these authors report that students with minimal geoscience knowledge as measured by the pretest (means of <40% correct) made the greatest conceptual gains in their geoscience courses. Students who scored higher on the pretest (40-60%) exhibited minimal changes in score, where as the highest scoring students exhibited no change. We observed a similar result in which students scoring less than ~40% on the pretest had the greatest gains, and students scoring 60% or higher demonstrated little gains.

Comparison between GCI results for our students and the national sample suggests that, overall, ESEE2 is no more effective than other introductory geoscience courses in increasing student knowledge of the geosciences. However, we note that students in sections 3-6 have overall learning gains on the GCI that are greater than gains for students in sections 1-2. Three out of the four sections, in fact, have statistically significant pretest to posttest gains of 7-8%. Additionally, for nearly all items common to both versions of the test, student learning as measured by the GCI was greater for students enrolled in these sections. We interpret these finding to indicate that the revisions we made to ESEE2 were effective in increasing overall student learning in this course. However, as our students are still achieving only 50% correct responses on the posttest, we clearly have a long way to go toward improved student learning.

Significant Changes on Individual GCI Items

A detailed look at the fraction of students responding correctly to each GCI item suggests that student conceptions changed for some of the course content but not for other content. Furthermore, statistically significant pretest to posttest learning gains was distributed among both easy and more difficult items; in other words, our students were not merely performing better on the easy GCI items on the posttest. What do we think we are doing right in this course? An analysis of student responses to each GCI item can reveal changes in student thinking, presumably as a result of instruction in this course. This type of item-by-item analysis of GCI data is also reported by Dahl et al. (2005).

The difficulties that students, as well as preservice and inservice teachers, have grasping the scale of geologic time and understanding the sequence of events in geologic history is well documented. Our students, however, had some of the greatest (and statistically significant) learning gains on items related to geologic time specifically item 28. Prior to instruction about 45% of students could place a series of events (life appears, dinosaurs appear, dinosaurs disappear, humans appear) in the correct sequence; however, only 25% of students could place these events in the correct sequence and at the correct locations on a geologic timescale. On the posttest, however, over 50% of students (and over 60% of students in sections 3-6) could answer this item correctly.

A detailed description of the classroom activity we use to teach geologic time is available at, presented as part of the 2007 Teacher Preparation workshop, In brief, the activity follows a modified learning cycle format in which students are presented with a "hook" to engage their interest in the subject, share prior knowledge about the subject, manipulate models or data related to the content, work collaboratively to reach conclusions based on evidence collected, and further reflect on the content learned in a homework assignment. This framework is used in all activities in this course. In this particular classroom activity, students first use an interview with an older adult to construct a timescale, which they divide into periods and compare timescales among the class. Next, students develop a geologic timescale of events provided by the instructor using only their prior knowledge. Students then use classroom and Internet resources to place the same events in the proper order and at the correct locations along the timescale. Finally, students investigate the geologic timescale and place aeons and eras of geologic time on the same scale as earth events. Comparisons are drawn between the human life timescale and geologic time.

Changes in student responses to item 28 suggests that this particular class activity seems to be effective in helping students to understand the nature of geologic time. Other classroom activities also result in high learning gains, as measured by the GCI. Our evidence suggests a loose correlation between type of learning task and conceptual knowledge change among our students.

During our course we spend at least one entire class period engaged in inquiry-based activities and group discussion of concepts directly related to geologic time (item 28), radioactive decay and absolute age dating techniques (items 21 and 1), plate tectonics and the structure of the earth (items 51 and 14), fossils (item 40), and groundwater (item 59). Concepts directly related to these GCI items are introduced to students through in-class activities and discussions, and reinforced through homework reading and quizzes. The greatest learning gains on the GCI were among these items; items 28, 40, 21, 59, and 51 had pretest to posttest learning gains of over 5% among all students tested, including some course sections with statistically significant gains on these items.

Topics related to earthquakes (item 70), the cause of ocean waves (item 34), the breakup of Pangea (item 73), and the location of glaciers (item 9) are discussed in the class but are not the main focus of the classroom activity. Some of the concepts related to these GCI items are covered in out-of-class reading assignments. Gains on these GCI items were lower than for the prior set of items, generally less than 5% (with the exception of item 70). Finally, concepts related to the past size of the earth (item 63), erosion rates (item 2) and the definition of a tectonic plate (item 6) are addressed in out-of-class homework reading assignments but class time is not spent discussing these concepts. All of these items show essentially no change or decreases between the pretest and posttest.

Our analysis suggests that our class format in which students learn content through hands-on activities supplemented by class discussion and homework reading leads to the greatest conceptual gains, as measured by the GCI. The impact of class discussion on conceptual learning is inconclusive; of the 3 GCI items that tested content learned only via class discussion in our course, two resulted in gains and one in essentially no change as measured by the GCI. Out-of-class reading appears to be the least effective means of student learning. Our data suggest that students need multiple opportunities in different formats, such as classroom activities, discussions, and supplemental reading materials, to develop a better understanding of geoscience concepts.

Persistence of Prior Knowledge and Entrenchment of Student Ideas - Libarkin and Anderson (2005) report extreme persistence of preconceived ideas despite instruction, as measured by the GCI. Their work, along with additional work, points to deeply entrenched ideas related to geologic time, events in geologic history, and absolute age dating. Our analysis of several GCI items, as exemplified by response patterns to item 1, also illustrates the persistence of prior knowledge.

Item 1, the most difficult item on the GCI, asks students to choose which method(s) scientists use to determine the age of the earth. This item is difficult conceptually, but also because students are prompted to select all responses that are correct (yet only "C" is correct). In our study, no students chose only the correct response (C) on either the pretest or on the posttest. On the pretest, only 20% of students selected analysis of uranium-lead as a method for determining the age of the earth, in addition to choosing additional distracters. On the posttest, however, over 45% of all students (and over 55% of students in sections 3-6) selected the correct response, suggesting that a substantial fraction of students did learn that the uranium-lead dating method can be used to determine the age of the earth. However, all of these students also selected other distracters in addition to the correct response on the posttest, suggesting that they held on to their prior conceptions that fossils, sedimentary layers, and carbon-14 dating were also methods used to determine the absolute age of the earth. It appears that we were successful in introducing the correct concept to our students, yet the students did not abandon their prior conceptions. This example points to the extreme entrenchment of prior ideas and to the challenge of altering these ideas during instruction.

The GCI has been used in our course as a means to gauge students' conceptual understanding prior to instruction, and to help us understand where our instruction has, and has not, been effective. We find that the population of preservice elementary teachers in our study has the same pre-instructional concept knowledge of the geosciences as other undergraduates, as measured by the GCI. This is in contrast to many studies that report weak background in science among preservice and inservice elementary teachers. Large pretest to posttest gains on some GCI items suggest that our course is particularly effective in altering certain prior conceptions, especially those related to geologic time. However, analysis of other items points to the extreme entrenchment of student ideas, despite instruction.

Furthermore, overall student performance on the GCI posttest remains poor, indicating that we have further work to do toward improving instruction and student learning in this course. Our analysis suggests that engendering conceptual development among students is difficult and that instructors may need to target certain concepts and spend more time on these. Class activities and group discussion, in particular, may also help students to develop scientifically correct conceptions. Out-of-class homework assignments appear to be least effective in bringing about conceptual understanding.

Educational Anomaly of Earth Science

The reputation of the geosciences as a "Rocks for Jocks" curriculum has long permeated K-12 districts and college campuses. In undergraduate programs, Geology has sometimes been viewed as a more accessible science requirement-satisfying option for non-science majors. On the secondary level, earth science has been a course traditionally offered to non-college-bound populations, as an alternative to the more "academic" track including Biology, Chemistry, and Physics. Despite the National Science Education Standards' call for Earth Science to be taught at each K-12 grade level (NRC, 1996), Earth Science is increasingly relegated to earlier grades, thus allowing college-bound high school students to take at least one Advanced Placement class in their senior year in lieu of any earth science at all. In California, for example, Earth Science is last taught as a comprehensive course in the sixth grade.

While elements of the geosciences are incorporated into later Integrated Science curricula, it is not atypical for California's youngsters to obtain their only exposure to the breadth of earth science in their first year of middle school: and this is in a state where earthquake preparedness is a required topic throughout one's public education. In New York State, where the Board of Regents has maintained statewide standards by requiring exit exams in all high school-level curricula, Earth Science has been relegated increasingly to middle schools. Earth Science is neither a topic for slow learners nor for young learners. This conclusion was never as apparent to these authors as during the course of an unintentional, unplanned "experiment," in which what is effectively a high school syllabus in Earth Science was taught to groups of upper division undergraduates at San Jose State University.

The Earth Systems Approach

Geology 103, Earth Systems Science, is a course that was developed by Prof. Ellen Metzger in 2000 at the request of the Director of the Program in Science Education. It was recommended that SJSU's pre-credential and inservice teachers would greatly benefit by having a comprehensive course in which earth science would be explored in depth. While the curriculum was initially designed to satisfy the needs of secondary science teachers, Geology 103 was recently mandated for all students enrolled in the university's multiple-subject credential program, since it satisfies content and pedagogy requirements as deemed necessary by the California Commission on Teacher Credentialing (CCTC) and the National Council for Accreditation of Teacher Education (NCATE). As such, the curriculum also awards General Education credit (thus fulfilling part of university-wide graduation requirements) to keep the prospective teachers' bachelor's degrees within an acceptable credit range.

The course was piloted in spring 2001, and has been taught since then by either Ellen Metzger or Paula Messina, both joint appointees of SJSU's Geology Department and Program in Science Education. Dr. Metzger brings to this course over a decade of experience as co-director of the Bay Area Earth Science Institute (BAESI), an SJSU-based program funded by the National Science Foundation, SJSU, Chevron Texaco Corporation, and other community partners, which has served over 1,100 Bay Area teachers in summer and Saturday workshops. Many of the BAESI teachers indicate that their motivation for participation correlates to the increasing expectation for them to teach Earth Science with little or no prior coursework in the subject. The opportunity to work with pre-college teachers has influenced Metzger's own teaching and she now incorporates more active learning strategies in her own classes, including Geology 103. Prior to her appointment at SJSU, Dr. Messina taught high school Earth Science to middle- and high-school students in New York City. For most of her twenty-year K-12 teaching career Dr. Messina taught the New York State Regents syllabus in Earth Science, which was offered in her school predominantly to "gifted" freshmen and to juniors who were deemed "under-qualified" to succeed in Chemistry.

A Tale of Two Credentials

As part of the science teacher credentialing process in California, one must demonstrate subject matter competency in the geosciences, which is typically accomplished by successful completion of courses in Astronomy, Geology, Meteorology, and Oceanography. Given the time-intensive professional demands of new teachers, it was clear to many of SJSU's Science Education faculty that requiring four semester-length courses was unrealistic, and that the salient content and process skills could be covered in a single comprehensive earth systems curriculum.

In contrast, New York State-licensed Earth Science teachers may obtain their teaching credentials by proving subject matter competency in a single discipline, geology for example, despite the fact that the Regents-endorsed syllabus incorporates atmospheric, oceanographic, and space sciences.

A more problematic corollary to the "demotion" of Earth Science to earlier grades is the less-comprehensive teacher science training requirements for those seeking multiple subject (elementary school) credentials. It is possible that someone teaching Earth Science to sixth-graders in California may never have taken a geoscience class at all. Hence, as his or her terminal earth science experience, a student may have a teacher who has little more content area and process skills training than any number of liberal arts college graduates. In New York State, one may obtain a high school teaching license specific to a discipline (i.e., Earth Science, Physics, Chemistry, etc.), or a middle school "General Science" license. When the high school-level Regents Earth Science curriculum is offered to eighth graders, as is done increasingly, there is no guarantee that the teacher of that course has successfully completed a single college-level course in any of the geosciences.

There appear to be as many science teacher credentialing standards as there are national, statewide, and district K-12 science content standards nationwide, and in no discipline is it more apparent than in the geosciences. The old adage that "anyone can teach Earth Science" (Nuhfer, 1990 is still routinely applied in schools around the country. The Council of Chief State School Officers (2001) reports that there are severe shortages of certified teachers in Earth Science, with only 82% of teachers across the nation being certified; certification rates in physics, chemistry, and biology are consistently higher. What a disservice we are doing to our youngsters by first portraying Earth Science as a "lesser" curriculum, not to be compared to the traditional college-bound science courses, and then allowing less-rigorously-prepared teachers to head those classes.

Given the current climate, it is not surprising that high school students opt for classes that carry more perceived academic clout (and in which the teachers may be better prepared), and that prospective science teachers consider careers instructing Chemistry or Physics - where they know they will encounter more motivated students. Thus is the self-fulfilling prophecy of the Earth Sciences in K-16 education.

A Single Course to Satisfy A Diverse Audience

In addition to the obvious streamlining benefits to K-12 teachers, Geology 103 was designed to incorporate effective pedagogy as an integral part of the curriculum. It is not a traditional instructor-centred lecture/lab class, but one in which students practice the process of science through activities, discussion, and guided inquiry. The advantage to K-12 teachers of this method of instruction to K-12 teachers is apparent, since they are offered a hands-on approach to complex concepts which they can then in turn adopt for their own classrooms. Class notes and activities have been made available online through the following Web sites so that teachers may freely print and use Geology 103 activities and slide sets with the kids they teach.

Among the benefits to non-teachers enrolled in the class is the opportunity to learn by doing, a method that has been all but overlooked in traditional college lecture courses. Non-teachers may enroll in this class to satisfy one of many General Education requirements for graduation, but the class is open only to juniors and seniors. By scheduling the course in the evening (to allow inservice teachers the opportunity to take it during the school year), Geology 103 has typically attracted non-traditional upper level undergraduates with majors as diverse as Marketing, Psychology, and Music in addition to those pursuing master's degrees in Education and Natural Science. Most people who have taken Geology 103 are working adults attending college part-time; so far, students have ranged in age from their early 20s to their late 70s, with a majority in their late 20s to mid 30s.

While in its planning stages the success of Geology 103 seemed to be a challenge, given its diverse audience. As an approved General Education course, over-arching goals included "the cultivation of knowledge of the scientific study of the physical universe and its life forms; an understanding and appreciation of the interrelationship of science and human beings to each other". As a course that would satisfy pedagogical- and subject area-competency requirements for both multiple- and single-subject credential teachers, it also needed to incorporate the scope and depth mandated by rigorous California state education standards. Hence, goals of the curriculum are numerous, and often dictated by external committees or agencies.

Assessment of student work includes exams, papers (there is a significant writing requirement for all SJSU GE courses), oral presentations, concept maps, and "WebQuests." A WebQuest is a type of guided Internet-based exercise developed by Bernie Dodge and Tom Marsh at San Diego State University. WebQuests consist of a task that is interesting and doable, learning advice, and links to Web sites selected to facilitate the task. Geology 103 WebQuests focus on controversial topics such as global warming and what killed the dinosaurs. Informal assessments permeate the course, especially during and after its student-centred hands-on components.

Despite the multiple and coincident requirements, Geology 103 has proven to be a popular course among teachers and non-teachers alike. The online component is widely appreciated by its enrollees who have professional and personal commitments beyond those of the traditional undergraduate, and its integrated hands-on design makes its content accessible to students of varied backgrounds and interests. Furthermore, the Earth Systems approach promotes the multidisciplinary nature of science, wherein an understanding of the whole requires an awareness of many components and how they interact, concepts sometimes overlooked within a narrower discipline-specific view of science. Other themes, which include scale, cycles, energy, and human interactions with Earth spheres, encourage students to find patterns, analyse scenarios, and construct solutions to pertinent everyday problems.

Questions of Suitability

In the fall 2001 semester, Paula Messina taught Geology 103 for the first time; after consulting with the curriculum developer and initial course instructor, Ellen Metzger, she adapted the course to suit the needs of its students (by reviewing previous course evaluations), while teaching to her strengths (which encompassed traditional geology and a broader physical geography/ Earth science background). The textbook selected, Earth Science, 10e. by Tarbuck and Lutgens, delves into the solid Earth, atmosphere, hydrosphere, biosphere, solar system and beyond, with careful regard to their influences on each other. Messina immediately recognised that the scope of the selected text was very similar to the Regents Earth Science course she had taught in New York State. Regents Earth Science was developed for high school students, and as mentioned previously, it is increasingly being offered to students in earlier grades. Despite the fact that Geology 103 is an introductory course (which carries General Education credit), it seemed impracticable to present the material in a comparable manner and at a similar level to adult learners. Would the students sense condescension if this were the case?

Messina pondered these dilemmas before the start of the semester, reflecting on experiences while leading summer institutes for BASEE (Bay Area Schools for Excellence in Education), an NSF/Hewlett Packard sponsored program designed to enhance elementary teachers' science content knowledge. It was through BASEE that Messina recognised the K-6 teachers' needs for basic understanding of scientific concepts. During two weeks of BASEE workshops in 1999, Messina developed and taught Earth Science content through a variety of hands-on methods to adult learners. While the teachers enrolled in the workshops had a true motivation to learn, higher educational achievements, and much more life experience than high school freshmen, their prior knowledge of the geosciences was somewhat similar. The course was taught at a level comparable to the NYS Regents Earth Science syllabus; many of the activities were taken directly from the labs developed for high school students. To some, this was the first such instruction they had ever received.

The BASEE teachers remarked that they were grateful to finally have had the opportunity to learn geoscience concepts (which they in turn were required to teach their own sixth-graders) in a way that they could understand. They were honoured to have been treated as adult learners, although admittedly learning content typically made available to much younger populations that they had never had the opportunity to explore previously.

Could it be that, when planning for Geology 103, one should assume a similar lack of prior knowledge? Would college students, particularly non-traditional upper-classmen, benefit from exposure to a syllabus that was at the same time being taught to middle- and high-school students 3000 miles away? Or would it be viewed as "dumbing down" the curriculum?

The "Experiment"

In fall 2001, the content and teaching methods of Geology 103 closely paralleled those of the Regents Earth Science syllabus in which Messina had ample experience. Each class began with a "hook," a discrepant event or pertinent demonstration of an unexpected phenomenon, a hands-on activity (conducted by an individual or groups of students), and a series of instructor-led discussions which were illustrated with PowerPoint slide sets and animations. Activities were adapted from labs Messina had conducted with her ninth-graders; language and procedures were modified somewhat to reflect the age and experience of adult learners. Tests were constructed with selected questions from prior New York State Regents Earth Science exams, sometimes verbatim.

Long before the publication of the AAAS's Science for All Americans New York State had an ongoing history of implementing subject matter standards in the sciences, as well as for all other secondary school subjects. Exit exams, administered by the New York State Board of Regents, are developed and administered annually as the terminal assessment instruments for standardised state-approved curricula. Earth Science is a Regents-level curriculum in New York State, an elegant amalgam of earth and space sciences. Two contributors to this chapter - Messina and Paul Speranza, a recently-retired New York State Earth Science teacher and co-author - have collectively taught over 5000 students in total, representing over 50 years of Earth Science teaching.

The Regents syllabus in Earth Science was first developed in 1970, but has been revised several times since then to reflect the dynamic nature of the geosciences. As with all NYS Regents science syllabi, the course requires at least 30 hours of laboratory work to be conducted by each student; those failing to meet this requirement are barred from taking the uniform exit exam. Earth Science was the first curriculum to include a "practical" component to the cumulative statewide exam, which tested students' laboratory skills (i.e., measuring, classifying, etc.) and reasoning abilities (i.e., extrapolating and interpolating student-collected data). This practical component had been long-touted as a "model instrument after which the exit exams of other curricula were to be re-designed. It is still the only Regents science with a performance portion of the exit exam.

The curriculum itself spans geology, oceanography, meteorology, and astronomy. As a basis for understanding many of the Earth systems explored, metric measurement, density, gravitation, energy conversions, transfer methods, and specific and latent heat (and their influence on adiabatic lapse rates, etc.), are also an integral part of the curse. In the college class equivalent, study guides were distributed before each exam and made available through the course Web site. Each study guide contained an outline of directed questions reflecting the scope of required content. Students had voted in favour of multiple-choice format questions for their first and second exams, and so each unit test contained items taken directly from New York State Regents exams (databases and old exams are available in the public domain from several regional school districts and New York State's Earth Science Regents archive). In each case, the study guide was used as the outline from which questions were selected.

As per the design of Regents-level multiple choice exams, selected questions required higher-order reasoning and an application of knowledge and process skills. Rarely, if ever, does the NYS Regents include questions that may be answered by simple memorisation of trivial facts.

Before Geology 103's first exam was administered, it was unclear whether the level of difficulty of the exam was appropriate for upper-division undergraduates; afterwards, it was a revelation to assess the results.

The test, which was composed of 50 multiple choice questions (four choices per question), was viewed as being "very difficult" when students were queried. One student spoke to Messina after class, and explained that "this is an introductory class," and that the exam was unnecessarily challenging. The comments were as shocking to the instructor as were the summative statistics: mean score (66%); range (38% - 94%). The second exam yielded similar results: mean score (74%); range (43% - 94%).

Before the third and final unit test (N.B.: unlike the Regents exam, the final for Geology 103 is not cumulative), many students requested greater variety in the types of questions included; some claimed that they historically blundered on multiple choice/short answer exams, and they preferred a short essay format. To satisfy a majority of students, the third exam was designed to include the student's choice of either 40 multiple choice questions (100%), or 30 multiple choice questions (60%) and 2 short essays (40%). All students were given the opportunity to select the 40 (or 30) short-answer questions of their choice (without restrictions, out of a total of 50); if opting to answer essays, students could choose two of a possible three.

Of 20 students who took this exam, only 6 chose the multiple choice/short essay format. Results for both formats were comparable: mean score (64% vs. 72%, for the multiple choice and mixed-format tests, respectively); range (41% - 97% vs. 33% - 95% for the multiple choice and mixed-format tests, respectively).

Assessment

What could these surprising results imply? Perhaps the quality of instruction was deficient, but perhaps the curriculum is far more difficult than it has been traditionally viewed. A comprehensive course evaluation was devised and distributed to students along with the "final" exam. In it, open-ended questions asked for students' opinions on the scope, sequence, and presentation of the material as well as the efficacy and methods of instruction. The results of the questionnaires were revealing, and reflective of enrollment diversity. Some students (non-science majors) took the course to satisfy a university-wide science requirement, while others (the current and future high school science teachers) either possessed or were near completion of a science B.S. degree. Although the comments were anonymous, the responses fell into two basic categories, likely reflecting this "dual-purpose" enrollment culture. While most students seemed to feel that the level of difficulty of the material was fair, a few remarked on their surprise at how challenging they found it - "this course was very difficult for me" - while one student admitted: "I felt that the material was just right for me, but I'm a grad student (Master's); I felt it may have been too detailed for undergrads."

Teachers and future teachers were asked whether they planned to use any of the in-class activities in their own classrooms, and every one admitted that they already had! Comments collected from just the non-teachers indicated that they unilaterally found the hands-on approach helpful ("I thought all the activities were useful and helpfulin applying course elements"; "I felt that the activities helped the subjects to come alive"; "It helped to learn the material by actually seeing it").

Students were unanimous in their appreciation for the online resources available to them; they found it helpful to print out the notes and activities. Similarly they found the instruction to be effective. Comments about the instructor included: "She made difficult material seem a lot easier than it could have been!"; "(She) kept the information given in digestible amounts"; and "She explained things simply, with real examples and references."

Comments about the level of difficulty of the exams included such statements as: "(They were) the most difficult exams of all my classes this semester"; "I found the exams to be very difficult"; and "They are very hard." The only comment which can be interpreted as being in some way different came from a current science teacher (who volunteered her identity on the questionnaire); she wrote: "I loved the exams! They made you think! While some questions were fairly straightforward, others made you take the information to another level of understanding. Great!"

Among the most flattering comments were from two non-science majors, one who stated that he or she was intending to take additional lab classes in geology, and one who admitted that he or she was contemplating changing majors, and becoming a high school science teacher.

To understand geologic phenomena, one must have the ability to envision three-dimensional processes, and must be able to imagine their progressions over vast amounts of time. This "fourth-dimensional" intellectual capacity is a fairly sophisticated hallmark of a formal operational stage of cognitive development. This level of abstract thought may first evolve in adolescents, but some individuals never achieve this level, even well into adulthood. Could the subject matter itself simply be too difficult for most adolescents, and many adults?

Why then do school districts around the country offer Earth Science to students whose cognitive development is characterised a s being at more of a "concrete operational" stage? The curriculum needs to be adapted to the linear-reasoning capacities of its audience, and often the topics are necessarily limited to descriptions, identification, or memorisation of material. When Earth Science is taught in this concrete-operational manner, it is indeed true to its reputation as Rocks for Jocks," and the status of Earth Science remains misleadingly low. Hence, when adults enroll in a course on the undergraduate level, their expectations may be unrealistic, and their success is somewhat dependent on whether they have achieved the cognitive skills to grasp abstract thought.

The Upside-Down Pyramid

Today's high school science sequence was developed over a century ago, when science was far more descriptive and concrete. Since the 1800s, it has been common practice for freshmen to take General Science or Biology, sophomores Chemistry, and juniors Physics. Geology hardly existed as a separate discipline at the time that this structure was first instituted, and now -even despite the plate tectonic revolution- it still remains as an elective at best, or more frequently, as an alternative to the traditional science menu for students of lower academic achievement. According to the National Centre for Educational Statistics (2000), in 1998 (the most recent year for which data are available), only 20.7 percent of high school graduates took Geology or Earth Science, compared to 92.7 percent who took Biology and 60.4 percent who took Chemistry. These are consistent with those of a survey conducted four years earlier, when 24.4 percent of high school students took a geology or Earth science course. If there is any significant trend indicated at all, Earth science courses are becoming less available on the high school level over time.

At the high school level, juniors and seniors who are deemed at a lower academic level were able to excel at the more conceptual topics (such as interpreting contour maps). The "brighter" freshmen had problems with the visualisation associated with this exercise. This antiquated progression has been the subject of a relatively recent movement to put Physics first. While only a few schools actually have retrained their teachers and made the difficult transition to ninth-grade Physics, there is increasing consensus that flipping the sequence gives the curriculum a coherence it now lacks, and allows students to build on concepts they have learned. By inverting the traditional succession, students would be given the tools to grasp complex concepts, and by delaying the fourth-dimensional concepts to the junior or senior year, they would more likely have already developed the cognitive skills characteristic of a more mature brain by then.

There is evidence that changing the sequence improves student achievement and interest in science. The number of students successfully completing Advanced Placement courses, and obtaining college credit while still in high school, may be tied directly to the shift in focus. At North Hunterdon High School in Annandale, N.J., for example, there were only 38 students enrolled in any Advanced Placement science class in 1990-91, the last year of the old sequence; the new curriculum has brought steady increases in those numbers, and in 1999, a record 226 students are in Advanced Placement science: 98 in Biology, 49 in Physics, 41 in Environmental Science and 38 in Chemistry.

While there has been a push toward the development of an Advanced Placement course in Earth Science, such a curriculum has not yet been approved. Perhaps we should consider completing the secondary science sequence with Earth Science as the "capstone" course. If students were to take such a course after having been introduced to requisite physical science concepts in Chemistry and Physics, and after learning about the diversity of life on Earth in Biology, Earth Systems Science may seem to be a natural finale, providing the student with a cohesive culmination of what may otherwise be viewed as discrete, disjointed disciplines.

The unplanned experiment in which a high-school level earth science curriculum was presented to college juniors and seniors is revealing in many ways. First, it confirms the difficulty of a science that has traditionally been viewed as the one of the least challenging options at the college level. Second, it validates that most people arrive at college with little or no prior knowledge in the geosciences, a failing of our current K-12 science requirements and sequencing. And third, it suggests that Earth Science may be best suited for pre-college students who have mastered physical/biological science foundations, and have attained the ability to reason and think abstractly. It is clear that Earth Science should not be relegated solely to earlier grades, or solely to under-achievers; it needs to be viewed as a culminating course, offering its students the ability to tie in prior knowledge with widely-observed everyday geoscience applications. The under-representation of Earth Science at the high school level is a reflection, in part, of the shortage of well-prepared teachers. Courses such as Geology 103 and professional development programs such as BAESI and BASEE are needed to help bridge the "preparation gap," and yet more needs to be done.

Raising the Level of Geoscience Awareness

We believe that there is a lack of geoscience awareness by the public including students and faculty on college campuses. Although the public is locally conversant with the general aspects of earthquakes, volcanoes, and dinosaurs, average citizens know little of the geosciences and the role they play in society. Important global issues of environmental degradation, climate change, and geologic resources rarely make headlines, but these are clearly important issues to the average citizen. Television programs such as NOVA and the specials on the Discovery Channel make the geosciences more accessible than ever to the public, and the popularity of such television programs demonstrates interest in science by the public. Despite this apparent accessibility and interest, the public largely lacks a general understanding of earth systems and processes.

In this chapter we summarise and analyse two college lecture series that address the issue of low-level public awareness and understanding of the geosciences. These lecture series draw large numbers of members of the community onto campus and highlight issues and current topics in the geosciences. We also present what we believe are some of the necessary ingredients to create a successful science lecture series and the benefits such a series can bring to your college, your community, and the geosciences. We write this summary having produced 11 lecture series (3 to 7 talks each) with over 40 individual talks over the last 5 years. We highlight strategies used at two different types of institutions on either side of the United States: one a community college in the resource-rich and hazard-prone Pacific Northwest, and the other a small very selective 200 year old liberal arts college in the tectonically stable Northeast US that has a host of environmental issues inherited from a long history of industry. The Spokane Community College lecture series is largely non-thematic and focuses on geology; whereas the series at Union College often follows a thematic framework and includes geology as well as environmental studies and allied topics. Given the great differences between our schools, we suggest that other departments who stage these events will enjoy similar benefits.

We realise that many geoscience departments run regular successful colloquium series that are largely geared to the scientific community on campus, or just the geoscientists on campus. There is clearly a need for such colloquia, but here we are advocating a different approach that emphasizes the art of communicating science to the public. How unusual is our approach? Our web searches determined that the number of programs that run such "outreach" lecture series is extremely low. It would appear that the traditional formula is most widely used in geoscience departments. As such, in this chapter we outline our approach and philosophy on running such a series, so that others who might be interested can benefit from our mistakes and lessons learned.

Benefits of a Geoscience Lecture Series

A successful lecture series takes an enormous amount of effort, but the faculty, students, campus, and community all benefit. Across campus, a lecture series could focus the attention of the campus community on specific issues. Such a series engages students and offers them an opportunity to meet with top scientists in their respected disciplines. Good speakers who can communicate the passion of science, the excitement of field-work, and the thrill of laboratory discoveries might result in increased enrollments. A high-profile series may also raise the status of a geoscience program on campus. College administrators strive for positive public attention and interaction with the community; getting the public on campus for positive events is one of the best ways to achieve this. A public lecture series may also sharpen and enhance relationships with the local business community. With corporate support, local industry is more apt to hire students and offer internships. Companies want to sponsor such a series because a modest expenditure can bring improved community relations. Finally, an earth science lecture series offers a wonderful opportunity for professional stimulation. Interacting with top scientists brings significant personal professional rewards for both faculty and students. Experiencing the wide-ranging disciplines of study and research may stimulate new ways to view our research and teaching.

Ingredients for a Successful Series

For five years we have held large campus-wide seminar series to highlight either geology or environmental studies. In putting these programs together, we independently came up with a number of guidelines that we use to assure that the series is well received by the campus and local community. Some of the points we review below seem obvious, but it is remarkable how often lectures fail to capture and hold a regular audience. Overall, it is important that the series is well organised and clearly laid out with fixed dates, times, and informative talk titles. We have found that there are several crucial factors that ultimately contribute to the success or failure of a public lecture series: (1) popular topics with broad appeal, (2) dynamic and enthusiastic speakers, (3) widespread campus and community promotion, and (4) organisation. With this seemingly simple recipe, a successful well-attended geoscience presentation can be achieved. Here we briefly comment on each point.

A Series with Broad Appeal: Exciting popular topics make a program successful. Our series have had a range of topics from catastrophic geologic events such as volcanic hazards, giant earthquakes, and massive floods to the origin of life, and snowball Earth. What may be interesting to you as a geoscientist may be very dull to the public. Regardless of the speaker's charisma, a talk on a complicated highly technical subject may not hold the interest of a general audience if not presented in a simple, understandable manner. Topics of local interest can be very successful because the topic may directly touch the lives of local residents. Consider topics such as wetlands, lakes, groundwater contamination, hurricanes, flooding, seismic activity, mining, petroleum exploration, radon, to name a few.

We also try to include talks that link science to human culture, public policy, politics, economics, or other mainstream disciplines. At Union College, we found that a mix of topics under a broad theme attracts a broader audience and gives students unique perspectives on a central theme. For example, a series entitled Water and the Law included a talk by a lawyer featured in "A Civil Action" who presented the legal aspects behind the Woburn MA, groundwater contamination case. His talk included how he had worked with geologists, and how the geology and the science behind groundwater were relevant to the case. For a series on Lakes and Environmental Change, a talk focused on the Lake Champlain basin, but showed artifacts imaged by side-scan sonar.

These artifacts included the missing boat of Benedict Arnold from the Revolutionary war, and the audience could clearly see the importance of the work in the context of American history. A final example is the politics revolving around the discovery of the Kennewick Man in the Pacific Northwest. Here the audience learned about the discovery as well as the political and cultural opposition to the research. Establishing a direct link behind the science and the human culture is important because for some, this is the first time they will see a connection of their own interests to science.

Multi-component Thematic Series: Another successful strategy is to present a series of similar topics or themes, as briefly mentioned above. A series theme allows linkage of seemingly unrelated topics and ties together different ideas surrounding multifaceted problems: ideal for an environmental studies program. A well-crafted and timely theme will almost automatically promote itself to the media. A series with an informal mission statement" is easier to promote to the media and easier to explain to the speakers. It is good to get this mission statement to the speakers early so they have time to adjust the content and direction of their talk. The approach here is somewhat different to how series are typically arranged in geoscience departments because we first determine the program theme and then find appropriate speakers.

At Spokane Community College we used a similar approach to address water quality issues in the Coeur d'Alene watershed. For this complex topic, we put together a panel discussion of professionals, scientists and governmental agency representatives from the many interests involved. Included on the panel was a biologist representing the local Coeur d'Alene tribe. We tried to give the panel balance, so the public could get exposure to all sides of the complicated scientific and social aspects of this diverse issue.

Media Contact and Advertisement: Once the program is arranged, the success of our series depends on proper advertising, public relations, and media coverage. Direct mailing of the series contents and press releases are crucial, and it is easiest to coordinate these actions with the campus public relations office. Press releases allow the series lineup to be published in the "Community Notes" or "Public Lectures" section of local and regional newspapers. When appropriate, we submit press releases for individual talks two days before the event to facilitate media coverage. If reporters do show up, we make sure that they have time alone with the speaker. For a speaker on the geochemistry of sediments in the soon-to-be-dredged Hudson River (NY), we coordinated post-talk press releases as well as interviews with the speaker. This took considerable coordination, as three TV stations and two newspapers covered the lecture because it came on the heels of a controversial EPA decision about dredging. Posters for the series are critical especially if they can be hung for the duration of the event ad month of more).

A head time of one month seems to be enough time to adequately promote an upcoming series. Over time we have developed mailing lists that target regional colleges and universities as well as governmental agencies, high school districts, high school science teachers, and individuals in the community. Announcements for flyers are written for the public, our principal target audience. These flyers have simple, short, and enticing abstracts. We fax announcements to the local radio and television stations as well as newspapers. Before the event, we offer interviews to the local talk radio stations if the topics are of specific global or regional interest. Finally, we create web pages for the lecture series with the series schedule and other details.

Links to Courses: We have had better success drawing an audience from the local community and less success with our own students. Several approaches seem to draw in some of our students. One is to tie the series or individual lectures to concurrent courses. We commonly find that speakers are more than willing to talk to a class during the day. At SCC, a talk on the 1991 eruption of Mount Pinatubo was organised to coincide with a Natural Disasters course offered on campus. Prior to the evening lecture, the speaker gave an afternoon presentation directly to the class on volcanic hazards and public awareness. At Union, the series Water and the Law was either required or strongly recommended to students who at the time were taking a course in either Groundwater Hydrology or Environmental Politics.

Information for the Speaker: Dynamic, enthusiastic speakers are what we want to ensure that the series will succeed. We make it clear to each speaker that they need to lecture to an audience that includes non-scientists; we refer to them as the "Nova" or "National Geographic" crowd. The ability of the speakers to reach the right level is absolutely crucial to the success of a program. We remind the speakers that they need to communicate their message to an audience with no expertise or knowledge in the subject area. All too often colloquium talks in science are so technical that even the students in the discipline are quickly lost, and as a result these students stop showing up to talks. In our experience, some speakers are better than others at delivering a simple message, and for many this approach might require new slides, overheads, etc. If the speakers and topics are not consistently interesting and entertaining, we expect our numbers to drop. Naturally, it is sometimes difficult to predict how a speaker will do, but it is crucial to inform them about the nature of the series and typical audience level.

Getting Established: We started small and through time our programs grew and developed a regular "base" audience. This base audience has been a good source for series-promotion in the community. For one early series at SCC, we began the season off with a presentation on the local aquifer and some of the controversial issues regarding the future of the aquifer. Because of the significant local interest in this topic, the crowd was large and in number of people that attended continued to s as large out such forums. A base audience of the general public is essential to future success, and initial high-profiled t seek geologic topics, if well promoted, consistently bring the public back to campus.

Honourarium, Funding, and Audience Discussion: It is important to provide an opportunity for the audience to meet the speaker and a reception following the talk seems to work best. Many in the audience have comments that they are not willing to ask during a question period following the talk, and many just want to share experiences. After a talk by a noted paleontologist at SCC, a number of young children used the post-talk gathering to collect autographs. As a rule, we provide the speaker with an honourarium and cover all expenses, which obviously vary from speaker to speaker.

Funding for these series can be obtained though departmental funds or financial sponsorship from local industry. Expenses for a speaker series include transportation and lodging, expenses, an honourarium, pre- or post-lecture dinner with fellow faculty, announcement flyers, and mailing costs. A series of four to six lectures per year can be done on a budget of \$2000 to \$4000 (US). At SCC, the idea is pitched to local mining exploration companies as well as geo-environmental engineering firms. These companies want a return on their investment, so it is made clear to them that we promote the company on the lecture programs and at the evening introductions. At SCC, a sponsor's board" is posted at the entrance to the auditorium, and the support is acknowledged at the start of each talk. Sponsor names are also posted on the series web page. In the Spokane area, many companies seem more than willing to contribute \$200 to \$500 per year.

Two Models of Success

The "Geology Lecture Series" at Spokane Community College. Spokane Community College, established in 1963, is located in Spokane, Washington, and is one of five regional colleges in the eastern Washington area. Spokane is a moderately sized city with a population of approximately 275,000, and the 12 month student population at SCC is c. 12,400. The Geology Lecture Series at SCC started in February of 1996 and has offered on average, four talks per year since its inception. Each lecture is free and open to the public and campus community.

The purpose for creating the series at SCC was to focus campus attention on the geological sciences, a discipline that had been underrepresented with respect to the campus liberal arts curriculum. The lecture series was conceived as a tool for drawing the attention of students, counsellors, and the administration to the importance of the geological sciences. To develop a viable funding base, a regional manager of a large exploration company was approached with the idea of a sponsorship of a geology lecture series. We soon had seven local companies who had agreed to contribute a combined total of $2,000.00. Since the initiation of this series, most of the funding has come from private sponsorship along with support from the SCC Student Activities Council.

Our first year focused on disaster-related topics: great earthquakes in the Pacific Northwest; volcanic hazards of the Cascades, the Missoula floods, and mass extinctions. The first presentation on earthquakes drew an audience of close to 125 people, and by the final talk in the series we had an audience of ~200. Several years later, the Geology Lecture Series averaged approximately 300 people per lecture with about 50% of the audience from the community. Over the last five years the series has focused on a diverse range of earth science topics and has not been thematic. In general we have found that a non-thematic approach has worked exceptionally well for our college. In part, we subscribe the success to the diverse population and its interests throughout the Spokane area. Since inception we have seen a significant drop in mining-related sponsors and an increase in geo-environmental sponsors.

In summary, the Spokane Community College Geology Lecture series has continued to grow in popularity with both the general public and college students because we have offered a diverse array of exciting and relevant geoscience topics. We have maintained a healthy sponsorship-funded budget and continue to actively promote the series through a wide variety of media, thus maintaining our dedicated base audience as well as continuing to reach out to the Spokane community. Union College is a small private liberal arts college in Schenectady (NY), which has a population of c. 62,000 and a campus community of 2,100 students. Union College was founded in 1795 and has since established strength in science and engineering. In collaboration with the Geology Department at Union College, the Environmental Studies Program has launched five Winter Seminar Series in the past five years.

These have been hosted in the evenings, generally on a set day of the week. The centrepiece of the campus is the 140 yr old Nott Memorial, a recently restored sixteen-sided stone building. We hold the seminar series in this building, which is not only a historical landmark, but is a well-know icon in the Capital district. This choice of venue helps raise the profile of the series, but creates some scheduling and security issues. Funding for the series has been derived from the ES operating budget, with occasional help from the administration for honouraria associated with high-profile speakers. These thematic series have included the following topics:

1. *Global Climate Change:* This first series included a number of Union faculty speakers who discussed the biology, geology, physics, economics and politics of global climate change. We chose this topic in the winter of 1997 because the Kyoto Protocol was fresh in people's minds, and it was an El Nino year. The total attendance at this series was only about 160 people.
2. *Water and the Law:* This series included three lectures aimed at exploring subsurface and surface water law and water rights. Speakers covered not only law, but also the science of groundwater, contamination, and land use practices. The lead off for this series was the lawyer behind the then-released "A Civil Action" who spoke to a capacity crowd. The total attendance at this series was approximately 700.
3. *Lakes and Environmental Change:* This series included seven speakers who covered topics such as the collapse of the Mayan civilization, the discovery of a sunken gunboat in Lake Champlain, and chemical stratification in Lake Nyos in Africa. As lakes are important for recreation in Upstate NY, this series had obvious regional appeal. The total attendance at this series was approximately 450.
4. *The Adirondacks:* This series featured five programs with a focus on the politics, geology, natural setting, and arts of the Adirondack Mountains. The total attendance at this series was approximately 650. This series was produced in collaboration with the non-profit Association for the Protection of the Adirondacks.
5. *The Hudson River:* This series featured six evening programs on the natural history, science, and culture surrounding the Hudson. The series coincided with intense controversy associated with proposed dredging of the General Electric-derived PCB's in the upper Hudson. Because the series coincided with the "public comment period" of the EPA's initial report, the debate was particularly engaged during the series, and we had excellent media attention as a result. The total attendance

at this series was approximately 1000. Like the previous series, this was produced as a partnership with the Association for the Protection of the Adirondacks.

A successful lecture series should bring stimulating, thought-provoking ideas to campus, and it should attempt to engage a wide audience. One fundamental measure that we use to judge success is attendance numbers and how those numbers change with time. Audience numbers give us a feel for how stimulating and accessible the talks are.

However, if the total number of people per presentation is consistently low, then the situation must be reevaluated. The demographics of the audience are also an important facet of the program and we note the percentage of: (a) students from around campus, (b) students in classes requiring the talk, (c) general community members, and (d) faculty and staff. Although high audience numbers are indeed a rewarding outcome, they are not the sole indicator of the series success. The fact that we continue to attract enthusiastic corporate sponsors for the SCC series indicates sustained interest.

We have outlined the collective attributes of our successful lecture series that grew over five years quite independently of one another. We found that it is important to run a well-organised series that is clearly laid out with well-articulated topics.

We ensure that the series has broad appeal across campus and within the community. To maximise exposure we contact media, write press releases, and advertise widely. Speakers need to be aware of the nature of the program, so we take time and explain the composition and expectations of the audience.

Multi-component thematic series are interesting to put together and allow a detailed examination of various angles on one subject. To increase student and faculty involvement, link the series or specific talks to concurrent courses.

Rewards may be significant for running a successful campus-wide lecture series. While these series take a considerable amount of effort, they tend to be stimulating and personally satisfying.

If the press or local businesses are involved, there is little question that a successful series will attract attention to your program on campus. At SCC, the geosciences have moved up the ladder in terms of enrollments, and resources. Much of this recent success can be attributed to the wide positive exposure the program has received from the lecture series. At Union, the seminar series has been successful in engaging the entire campus and it has been regarded as a campus model for how series should function.

The Environmental Studies program has received wide support as a result of this exposure.

In the end, we hope that the audience members from the community have become more scientifically literate. A high-profile successful series with well-known speakers and large audience participation will undoubtedly catch the eye of administrators who may be asking where new resources should be allocated.

3

Models of Inquiry-based Science Outreach

We live in an increasingly urban world. In fact, nearly half of all people, globally, currently live in cities. By 2030 this is projected to increase to nearly 60%. In the United States 80% of the population now resides in urban areas. Despite the large numbers of students educated in urban schools, only recently have geoscientists begun to teach students in urban areas about the geoscience issues affecting their communities. This is unfortunate because urban students with fewer experiences with geoscience may assume it is less relevant to their lives. As urbanisation continues, studies are needed that detail effective strategies for teaching geoscience in an urban context.

Efforts to improve urban geoscience education are also essential to enhancing the diversity of the feosciences. Demographic data reveal a clear disparity between the representation of many ethnic groups within the U.S. population and within the geosciences profession. For example, African-Americans compose roughly 12% of the U.S. population, but only 1% of employed geoscientists with a doctoral degree. Similarly, Hispanics make up roughly 12% of the population though only 2% of Ph.D. geoscientists. Over 85% of geoscientists currently employed with a Ph.D. in the U.S. are White, non-Hispanic. This pattern of under-representation is also apparent at the undergraduate level. A primary way to increase the involvement of minority groups in the geosciences is to provide experiences with inquiry-based geoscience education during elementary and secondary school. Studies have shown that students with positive early experiences with science are more likely to continue in science. As approximately 91% of minorities in the U.S. reside in urban areas, efforts to enhance geoscience diversity necessitate increasing effectiveness of educational strategies in urban settings.

At the Paleontological Research Institution (PRI) we have concentrated on providing urban students with access to inquiry-based science experiences through museum educational outreach. In this chapter, we describe three primary obstacles to improving urban geoscience education and then discuss two of the programs we have developed that have been offered to urban elementary school classrooms in central New York. We describe the goals and structure for each program and present evaluation data gathered from participants. Our findings suggest four best practices that have been integral to our programs and which can be employed in other urban outreach efforts.

Obstacles to Urban Geoscience Education

A variety of challenges exist to engaging students from urban school districts in the geosciences. These include: 1) the perception that urban environments are unnatural; 2) financial limitations in lower income urban school districts; and 3) limited English proficiency for a significant number of students in urban classrooms.

Perception of urban environments as unnatural - The misconception that the geosciences, and natural sciences more broadly, are less relevant in urban settings pervades popular thinking and poses a challenge to urban geoscience education. This sentiment is rooted in historical definitions of nature that have tended to restrict the study of natural phenomena to 'wild' areas and other rural settings. Some teachers with whom we have partnered have verbalised their belief that urban students live in greater isolation from natural phenomena than rural students, who they perceive to be surrounded' by nature. While urban environments may pose unique challenges to some traditional field-based geoscience activities, these are not insurmountable. For example, Haywick (2002) and Wetzel (2002) note the general paucity of accessible rock outcrops within urban areas but provide suggestions for alternative approaches. Many geoscience phenomena are manifested in urban settings. While some of these have been used in construction or landscaping, materials such as rocks, soils, and gardens as well as urban floras and faunas can all be used creatively in urban geoscience activities.

Limited financial resources - Financial limitations in some urban school districts pose a challenge to providing students with an adequate education. The case of The Campaign for Fiscal Equity vs. New York City (2003), for example, ruled that students in the New York metropolitan area were not receiving their constitutionally mandated basic education. While per-pupil spending differences between inner city and suburban schools have been shown to vary among metropolitan areas, inner city schools typically have older infrastructure, larger class sizes, less experienced teachers, and fewer educational resources. When per-pupil spending is adjusted for a variety of educational costs (e.g., educating non-native English speakers), school districts that serve areas with high-poverty consistently spend less per-pupil than those located in low-poverty areas. In other words, students in many urban public schools have more limited access to school resources and perhaps fewer opportunities for scientific enrichment than their peers in other school districts.

In our experience, outreach program fees are covered in a variety of ways: grant support, parent-teacher organisations, individual teachers or parents, and by the school district. In a lower income community, far fewer external sources of financial support are typically available for science enrichment programs. Limited financial resources may pose a primary obstacle to participation in some in-school and extracurricular enrichment activities that explore the natural environment (e.g., field trips).

Limited English proficiency - In many urban schools, classrooms contain a significant proportion of students who do not possess certain basic English reading and writing skills. This is due to a variety of factors, including the presence of students who are learning English as a second language. Within the public school system in New York State there are approximately 200,000 students considered to have limited English proficiency; these students come from over 135 different language backgrounds. Effective education of students with limited English proficiency requires additional resources, (e.g., teachers, tutors, books), yet these resources are not often accounted for in school budgets. In our experience, it is not uncommon to visit classrooms in which recently immigrated students are combined with native English speakers in the same classroom without obvious sources of language support. Beyond financial challenges, limited English proficiency poses an obstacle to conventional methods in science education which rely heavily on assimilating information through lecture and textbook learning.

These three obstacles may be overcome, in part, through museum educational outreach. The development of grant-supported outreach programs can reduce financial obstacles to participation by bringing educational resources directly to participants. Through the incorporation of local examples, outreach programs can also help to eliminate misconceptions about the nature of urban environments and help students to connect geoscience classroom content to their lives. Furthermore, project-based outreach programs provide opportunities for students of a wide variety of learning styles and backgrounds to participate in inquiry processes.

At the PRI, we provide educational programming to students from a broad range of socioeconomic backgrounds. These programs draw on students' intrinsic interest in fossils as a starting point for covering scientific processes and content across their curricula. Our programs are largely underwritten by state, federal, and corporate grant support, allowing us to alleviate some of the financial hurdles to participation and engage students that may not otherwise have access to such activities.

Demographics of schools served by the PRI outreach programs - Since 1999, we have offered outreach programming to schools located within an approximately 160 km radius of Ithaca, New York, with the majority of programs concentrated in the Ithaca, Syracuse, and Utica city schools and surrounding rural and suburban districts. Syracuse and Utica are two of the largest urban areas within central New York, with metropolitan populations of approximately 732,000 and 300,000, respectively. In our outreach area, the number of students eligible for free and reduced-price lunch (a proxy for family income) is generally greatest among urban schools, reaching up to 75% of the student population in some instances.

Our outreach area includes urban schools where 5 to 9% of students are considered to have limited English proficiency. Between 1999 and 2003 we offered approximately 500 programs at over 100 schools, involving over 5,000 students; roughly half of the programs were offered in urban schools. These programs were developed for a wide range of educational levels (K-12), though most were offered to elementary school classes. The Collections Connections and Devonian Seas programs were developed explicitly with urban classrooms in mind and are discussed here.

Collections Connections - The Collections Connections program was designed to teach urban elementary school students about natural history through the process of creating, interpreting, and exhibiting a collection of natural history objects gathered from around their neighbourhood. Creating a classroom museum from collected objects is an effective pedagogical strategy as it increases student awareness of their local environment, uses familiar and inexpensive objects as a basis for science education, exposes students to a variety of learning styles, and links science to other aspects of the curriculum. Collections Connections was pilot tested in fifth grade classrooms at Beverly J. Martin Elementary School, located in downtown Ithaca. The program involved 2 fifth grade classrooms each school year from 1998-2000 and was supported though a grant from the Shell Foundation. The grant enabled the PRI to offer an extended series of about 11 sessions for each class, and to bring the classes to the PRI for a tour of the exhibits and collections.

Classes participating in Collections Connections were introduced to the program through a field trip to the PRI and an interactive presentation on natural history and museums. After this introduction students went on a field trip to a small (1 square city block) urban park located a few blocks from Beverly J. Martin Elementary School. During this session, students collected and documented natural history specimens (e.g., rocks, twigs, and leaves) found in the park and along the sidewalks in transit to and from their school. A specimen label that contained locality information was created by the students for each object they found. This portion of the program emphasized the development of student observational skills through the process of collecting and studying natural objects. Student collections then served as a jumping off point for teaching about classification and mapping during subsequent sessions.

Students were asked to classify objects according to a system of their choice (e.g., texture, shape, organic/inorganic), and then contrast their classification scheme with those generally employed by scientists. Students then mapped their objects on a local city map and were asked to discuss why the location of their specimens was an essential piece of information for the interpretation of the nature of their neighbourhood. Once classified, students' specimens were the source of discussions during follow-up sessions on regional geologic history, Earth processes such as weathering, and the natural urban environment. Students invariably also collected pieces of man made materials (e.g., asphalt and brick) during their field trip. Rather than invalidating these finds as somehow outside the focus of natural history, these were instead incorporated into a discussion of the natural origins of man made objects and their place in the natural history of their neighbourhood. Several sessions were devoted to discussing the scientific meaning of their specimens and how they might be used within the context of a classroom museum exhibit to tell the story of the local environment.

In subsequent sessions, students worked in small teams to write exhibit text, illustrate their specimens, and interpret their findings for younger student visitors who took classroom museum 'tours.' During this process, students elected to participate in certain areas in which they had interest and strength, thus engaging them in activities in which they had confidence. They could elect to work as curators, exhibit designers, or exhibit developers. For example, a student with strong organisational skills might choose to play a role as curator within the classroom museum, a student with strong written communication skills may write labels as an exhibit developer, and a classmate who enjoys drawing might create artwork as an exhibit designer. Communication was especially important during this portion of the program, and could be accomplished in a variety of different ways (i.e., through drawing, writing, or speaking). This format provided opportunities for participation by all students including those with limited English proficiency who could play important roles in the exhibit "teams" as curators and exhibit designers.

In the first year, the culmination of the project was creating a virtual museum exhibit of their classroom specimens with digital images and interpretive text. In the second year, capstone experiences included field trips to the PRI and then to the American Museum of Natural History in New York City. These field trips served to reinforce important concepts about museums and collections, and provided students with additional contact with natural history specimens (e.g., examining the skull of a Tyrannosaurus rex, and searching for fossils in local Devonian-age shale).

Collection Connections was developed to be an inexpensive educational experience that drew upon local resources and trained teachers to undertake natural history projects in their surrounding urban environment. The project was intended, in part, to make teachers themselves aware that the varieties of building stone, gravel, cement and asphalt around them each have geoscience significance and can be connected to an inquiry-based curriculum. Moreover, if teachers run the project for several years their classroom will gradually accrue its own set of interpreted rocks and related objects that can be used in classroom activities. Most significantly, teachers who begin to look for the nature in their neighbourhood may begin to integrate other urban geoscience observations such as landscapes, weather patterns, and other phenomena into their teaching.

Devonian Seas - Like Collections Connections, the Devonian Seas project was designed to engage students directly in scientific processes such as classification and hypothesis testing. Devonian Seas differed, however, in its primary focus. Where Collections Connections took students through the process of collecting and interpreting natural history specimens, Devonian Seas was designed to engage students in authentic paleontological research using specimens collected by teachers during educator workshops led by the PRI. Such collaboration provides students and teachers with valuable inquiry-based experiences and scientists with important paleontological data. Involving students at all levels in authentic research is a critical component of many recent educational reform recommendations. For example, the National Research Council recommended that "for students to develop the abilities that characterise science as inquiry, they must actively participate in scientific investigations." Research experiences can play a powerful role in shaping student interest in science, but nave traditionally been restricted to undergraduate and high school students who have demonstrated an aptitude for science.

Devonian Seas was initiated in spring 1999 with the goal of engaging fourth through ninth grade students in research on the Devonian marine fossils of central New York. Since 2001, this partnership has involved approximately 100 classrooms per year in Syracuse, Utica, and surrounding rural schools. Approximately 3,000 students have participated to date, with roughly 50% of these from urban schools. Devonian Seas has been supported by grants from the National Science Foundation Geoscience Education program and the GE Fund, which have minimised the cost incurred by participating schools. A nominal mileage reimbursement payable to the PRI ($40-85 depending on travel distance) was charged per day of educational programming at each school.

For some teachers, involvement in Devonian Seas began with a 3-5 day workshop offered during the summer for which participants received in-service credit through the local school district. Field-based workshops served four primary purposes: 1) to introduce teachers to inquiry-based approaches in geoscience education; 2) to provide direct experience with the research questions and methods of the Devonian

Seas project; 3) to collect fossil materials otherwise unavailable in urban settings; and 4) to engage teachers directly in geoscience research. During these workshops, teachers and staff from the PRI collected fossil samples for use in classroom research, created measured stratigraphie sections, participated in Devonian Seas classroom activities, and discussed inquiry-based science pedagogy.

Student involvement in the project began with a hands-on classroom program led by a scientist-educator from the PRI. These programs were offered each school year between October and June and were generally 60 minutes in length. During this introduction the students' role as scientists within the partnership was made clear. Students participated in reconstructing the geologic history of central New York through examination of sedimentary rock samples and fossil specimens from the PRI teaching collection. By making observations and proposing hypotheses within a guided-inquiry setting, students became familiar with the geologic history of the region and the inquiry-processes they would employ later during their research.

The next phase of the project consisted of classes collecting data on the abundance of different fossil taxa found in a bulk sample of Devonian marine sedimentary rock. Teachers generally facilitated this classroom-based research portion of the project. Small groups of students worked together to identify and count fossil specimens using identification sheets. Students recorded their data and then analysed and interpreted their findings with the guidance of their teacher and the PRI scientist-educator. Discussions of their data informed student understanding of the geologic history of New York and of past biotic and abiotic change. On average classroom research took an hour, though several classes were known to extend data collection over several weeks of occasional 30-60 minute sessions. Completed data sheets were submitted in hard copy format to the PRI along with the fossil samples, student questionnaires, and post-program evaluations from participating teachers; an online form is in development that will allow students to submit their data electronically and see their contributions in a broader context.

Devonian Seas was designed to be accessible to different learners through its emphasis on hands-on activities involving the comparison and interpretation of fossil forms and through the use of image-rich educational materials. Providing students with authentic fossil samples on which to conduct research in the classroom also circumvented challenges associated with school field trips while still offering students opportunities for first-hand exploration of geoscience materials. Small-group research collaborations allowed each student to play an active role in aspects of data collection and analysis.

Evaluation of Outreach Programming

Both Collections Connections and Devonian Seas have been evaluated by participating teachers and students using internal and external evaluation tools. Internal evaluations employed post-program, semi-quantitative questionnaires completed by participating teachers as well as surveys with more focused questions about program content, activities, and areas for improvement. External evaluations of Devonian Seas were conducted by Maria Lawrence and Molly Loomis of the Program Evaluation and Research Group (PERG) at Lesley University and Stephanie Thompson of Seavoss Associates. External evaluations were administered in both rural and urban schools.

Collections Connections - Collections Connections was evaluated in three ways at the end of each year through: T) teacher surveys; 2) student surveys; and 3) student feedback gathered by teachers in open-ended classroom discussion. In addition, the PRI outreach educator (Ross) kept a journal during pilot tests and met with the teachers to discuss his observations and impressions. The evaluation focused on qualities such as student attitudes, interests, and awareness, in keeping with the overall goals of the series.

In the surveys, students were asked what they liked most and least, and what they found to be too difficult. Many students reported that they most liked the programs in which they created labels for specimens and made the exhibit, or best liked the activity to collect specimens in the local park. They liked least discussions of the science behind the objects, which students perceived to be overly lengthy. This suggests to us the need for more interactive and tactile approaches for these sessions. The process of creating the labels and museum exhibits required considerable coordination by the teacher and the PRI educator and necessitated teamwork among the students. While teachers and the PRI educator observed that this process was at times chaotic while students developed a system for working together, students indicated that they enjoyed the freedom to work independently from their teacher that led to an accomplishment that was readily seen by all (the exhibits). Few students reported that any aspect of the activity was difficult, but among those that did, the "rocks" part (identifying and interpreting the history of the rocks in the collection) was considered more difficult than other topics.

Personal observations made by the PRI outreach educator (Ross) revealed that students took a special interest in display of specimens that they had personally collected and frequently pointed out the particular specimens in the collection and exhibits that they had found. Teachers anecdotally reported that some students who were not otherwise academically active became more engaged with certain activities such as collecting objects or setting up exhibits. Many of the students had been to an art museum on a school field trip, but few had been to a natural history museum other than the PRI. The teachers frequently remarked to the PRI outreach educator during the series that most of the students in the class would not otherwise have any firsthand contact with museums or scientists.

Devonian Seas: Devonian Seas was evaluated through: 1) student and teacher surveys; 2) classroom observation by external evaluators; and 3) student and teacher interviews/small-group discussions with external evaluators. Participating teachers have indicated in the interviews and post-program surveys conducted by PERG and Seavoss Associates that the program satisfies their curricular goals and that their students enjoy and are actively engaged in the project. One teacher commented during a post-program classroom interview: "I thought it was excellent. It was so good in so many ways it gave kids a very good idea of it paleontological research, and it made it real for them". Lawrence (2001) compared and contrasted the educational outcomes of Devonian Seas with other classroom outreach programs (e.g., Dinosaur Science) offered by the PRI, and round that participation in authentic research in the context of Devonian Seas engages students in inquiry-processes in a way that goes beyond other more traditional hands-on classroom programs. Thompson also documented similar sentiments about authentic inquiry and quoted a teacher as saying that "answers are not always clear-cut," and that participation in Devonian Seas helped students to understand this. In addition, both Lawrence (2001) and Thompson (2003) found that students participating in Devonian Seas were aware that they were gathering real scientific data in collaboration with the PRI and took that responsibility very seriously. One fourth grade participant commented on a post-program survey, "there was a greater level of trust in us than I expected," while other students noted that this project was different from other classroom activities because this was real research and they (the students) were real scientists. In an illustration of student awareness, Lawrence (personal communication) reported that a fifth grade student whose data sheet had been overlooked accidentally, told their teacher several times, 'they need my data,' until it was mailed to the PRI.

Collections Connections: The teacher evaluations suggest that future series might focus on a narrower topic. One of the challenges of working with a classroom only once or twice a week is that each session needs to devote time to reviewing material so that all students start each class engaged and with comparable background. Thus, it is important to work with teachers to promote ongoing maintenance of student interest and knowledge throughout the course of the program. Additionally, teachers perceived a need for more examples from which students can work, ranging from how to identify rocks to how to write the text for labels. Teachers seemed more concerned about themselves and their students knowing vocabulary, correct Categorisation, and factual information than was intended by the PRI educators. In fact, it became evident that a future goal should be to increase the awareness among participating teachers that natural history in general, and geology in particular are inquiry-based approaches linked to many other areas of human inquiry and not simply collections of facts and names to be memorised. It is important that teachers work collaboratively through the project, and that teachers receive opportunities for professional development before the series begins.

Devonian Seas: Primary areas for improvement within Devonian Seas centre around improving project materials (e.g., identification keys) and developing ways to provide feedback to students after data collection has been completed. The Devonian Seas project was designed to involve a relatively minimal time commitment on the part of participating classes, to allow it to be integrated with full class curricula, and adopted by large numbers of classes. Now the challenge for the PRI educators is to provide additional opportunities for students to work with data and learn about the results of their research after they complete the required components of the project. Some students expressed frustration working with rock samples that contained small, fragmented, or sparse fossils. Future classroom presentations by the PRI educators need to clearly outline some of the challenges of working with fossil materials that are poorly preserved (e.g., not being able to identify all fossils in a sample), in part, to adjust student expectations of the research process. The introductory program should make it clear to students that scientists also find it challenging to identify fragmentary fossils and at times are unable to gather data from every partially preserved specimen.

Evaluations of Collections Connections and Devonian Seas have highlighted four pedagogical best practices for programming in urban schools that are closely aligned with recommendations from American Association for the Advancement of Science (1993), National Research Council (1996), and Barstow and Geary (2001). These best practices may be particularly effective in urban classrooms as they validate student contributions and background knowledge, draw on interdisciplinary topics of interest, and provide students with access to experiences they might not otherwise have had.

1. Access to authentic materials — Providing students with opportunities to touch and observe real fossils to which they may not otherwise have exposure can be a powerful catalyst for learning. The comments of one fourth grade student are representative of much of the feedback we receive from program participants. "Thank you for coming to our classroom. We all loved the fossils especially the shark tooth and the wooly mammoth. I think paleontologists are cool people because they can discover new animals and plants." Hands-on experiences with real fossils can engage students who might not otherwise be active classroom participants by drawing upon their interests and by entrusting them to handle materials of scientific and. cultural value. In addition, a hands-on approach provides opportunities for students of different learning styles, including those with limited English proficiency, to actively engage in scientific processes and content.
2. Use local examples — Using local examples connects classroom curricula to students' lives in a way that textbooks often do not. Incorporating local examples into classroom activities increases student confidence as it draws upon student experiences and background knowledge while demonstrating the connection between classroom curricula and student experience. This is reflected in the following fifth grade Devonian Seas participant's comments about the geologic history of their city: "I did not know that Syracuse was (once) an ocean with warm water. I thought it was always land

before. I also did not know there was a difference in shells today than when Syracuse was an ocean." Drawings created by Collections Connections participants also demonstrate an investment and interest in the objects they personally collected and described and in the natural history of their urban neighbourhood.

3. Incorporate interdisciplinary approaches and topics of interest — Using interdisciplinary approaches and topics of interest can be a springboard for teaching across curricula. Students who may picture science stereotypically as involving white coats and lab research, may feel more engaged when they learn that science can include approaches such as art, writing, and fieldwork. In Collections Connections, for example, students go into the 'field' and create their own collections of natural history specimens for which they later create interpretive exhibit labels and graphics. Exciting content, such as paleontology, that is interdisciplinary in nature, can also be used to reach a variety of content and process standards. For example, one fourth grade student commented, "I don't really like science or rocks, but fossils are my thing so this was great!" Furthermore, local examples can serve as an excellent means for meeting cross-curricular content standards. Students in the PRI outreach programs made connections between science and social studies/regional history through their interpretation of collected materials such as asphalt in Collections Connections and their reconstruction of the ancient marine environments of New York in Devonian Seas.

4. Engaging in research — Participation in authentic research teaches students that science is something that they can do. This type of experience can empower students to take an active role in their learning as reflected in this quote from a fifth grade student: "The coolest part of the exercise was that we were the scientists uncovering the evidence that we used to classify them." Another 5th grade student commented that the best part of the activity was "looking at the fossils because it gave me a head start on my career." Research collaborations involving scientists and precollege classes help to teach students about inquiry through direct hypothesis testing and data collection, and provide students with scientific role models. Students feel entrusted within such research partnerships and take their role as scientific collaborators seriously.

We must be committed to improving urban geoscience education in order to teach students about the environmental issues facing their communities, increase diversity within the geosciences, and provide equitable access to high-quality, inquiry-based science education. The Collections Connections and Devonian Seas programs provide two effective models for engaging urban students in the geosciences through museum educational outreach.

These programs created opportunities for students to take ownership and responsibility for tangible, authentic, and exciting geoscience projects. Both outreach programs were hands-on and inquiry-based. Both were also structured around students developing their own questions and hypotheses, and interpreting data individually and in small groups. These programs were also unified in their focus on engaging students in observation of regional and local natural history materials. Given the similarities between these two programs the question arises as to whether there are certain contexts in which one approach might be preferred over the other.

Though similar in goal, the two outreach approaches differ in their structure, content, and associated activities. Collections Connections was designed as a series, which featured longer-term interaction with the PRI educator and more in-depth inquiry and content. This format was more time-intensive for both the PRI educator as well as the participating classes, which meant that relatively few classrooms were able to participate.

In contrast, Devonian Seas was designed to be a relatively small time commitment for both classroom teachers and the PRI educators. This promoted the involvement of large numbers of students by facilitating easy integration into an already full curriculum and limited museum educator hours. Such a format necessitates the development of innovative ways to follow up with participating classes to provide feedback on student data and project research.

Collections Connections, through its focus on involving students in all aspects of collection creation and exhibition, connected students closely with their local urban environment and engaged students in using a diverse array of skills applied to understanding and communicating natural history. Our experiences show that students in urban neighbourhoods can find objects for a natural history collection and will take personal interest in them. The identity of the specific objects in this case is less important than the process of learning about the objects and sharing that information with others. In Devonian Seas, students were also involved in finding objects by using materials provided by the PRI. Devonian Seas participants came to see themselves as scientists through their direct engagement in research and collaboration with the PRI staff.

Perhaps future work could investigate ways to integrate these two approaches into a multi-stepped program for classes willing to commit a greater portion of their school year. Engagement in paleontological research could be undertaken first, after which students could create collections from their own neighbourhood that would then be interpreted using the research approaches with which they had gained familiarity. In closing, the following overarching point must be made: while this issue of the Journal of Geoscience Education is focused on ways to improve urban geoscience education, many of the approaches presented here are not only suited for urban classrooms. In fact, the best practices discussed here are broad guidelines for engaging any audience in scientific inquiry.

Teaching and Learning Experience

The importance of hands-on activities in learning science at all grade levels has been widely recognised for several decades from the results of a number of studies, including: 1. the development of constructivist learning theory, 2. the introduction of process-based approaches to science education (AAAS) and 3. the development of "discovery lessons" by the Science Improvement Study (SCIS) at UC Berkeley. Each approach has supported the merits of hands on learning in the sciences. However, multiple factors act as barriers to the full scale adaptation of hands on activities in classroom science lessons including: limited funds for materials, lack of experiences in teacher training and professional development programs, and limited adoption choices for quality resources such as textbooks and kit based instructional materials.

Across the disciplines of mathematics and science, widespread attention has been paid to the need for high quality training for pre-service teachers, particularly in the implementation of new teaching strategies. For instance, Brown and colleagues (1990) cite the need to shift focus to place greater attention on pre-service teacher training, which we agree is particularly important given reports that some in-service teachers who serve as mentors to young teachers resist implementing reform based pedagogy, and given the tendency for pre-service teachers to "revert to teaching styles similar to those their own teachers used".

Early teaching experiences for pre-service teachers have been shown to empower students to have confidence to teach, help them clarify their decision to become teachers (or not), and give them improved attitudes towards subsequent coursework. Similarly, early teaching experiences that include service learning has been shown to promote improved attitude in pre-service teachers as have programs that use inquiry-based learning environments to train pre-service teachers, both of which are components of the Hands On Lab (HOL) training for pre-service science teachers.

Need for Pre-Service Teaching Experiences

Previous research, state, and national standards have documented the need for pre-service teaching experiences for students interested in secondary mathematics and sciences. Additional work has indicated that early career science teachers often find short comings in their preparation in areas such as classroom management, content knowledge, pedagogical knowledge, and cite a "gap" between their university preparation and professional practices. Even students who demonstrate good content knowledge have reported negative attitudes towards science and science teaching, making a clear case for pre-service teachers to have low-risk, non-threatening teaching experiences early in their preparation. The Internship in Science Teaching course in the HOL provides such experiences for the undergraduate students (referred to as "HOL Interns").

Recognising the educational importance of hands on science activities as well as common barriers to their implementation in the classroom, a course was developed at California State University, Chico to meet the needs of pre-service teachers. This upper division course, Internship in Science Teaching, is taught in a dedicated laboratory, the HOL, which provides opportunities for undergraduate students to practice teaching hands on, inquiry based science activities to classes of elementary aged students. Classroom teachers bring their students to campus for the science lesson field trip and pre-service teachers learn the art of teaching science to these young students through supervised and structured hands on teaching experiences.

The State of California began a testing program in science in 2005 covering the 4th and 5th grade content standards. Recent changes in the credentialing process for elementary level teachers mandate that some level of mastery in science be attained and demonstrated by testing. However, pre course surveys of the undergraduates registered for the HOL course indicate that these students rate themselves low in scientific ability and are hesitant to teach science.

Compounding the problem of low confidence in science reported by undergraduates planning to be elementary school teachers, in-service elementary school teachers report feeling overwhelmed by the breadth of science content they are expected to know and teach. Elementary school teachers also struggle to find time to include science in daily or weekly curriculum due to the emphasis placed on mathematics and language arts; some schools have mandated periods of time set aside for math and language arts, but generally, dedicated time is not allocated to science in elementary grades. Better preparation in science and teaching may alleviate some of these concerns, potentially even improving retention rates of science teachers.

Research in mathematics education reports that pre-service teachers are often aware of teaching strategies outside their own learning experiences but do not feel well prepared to implement new teaching strategies, leading nearly 2/3 of surveyed students to question the feasibility of implementing reform teaching methods in mathematics.

In response to survey feedback, Frykholm (1999) points out the need to provide teaching experiences that implement new teaching strategies during undergraduate coursework. Such is the case with the HOL where students are provided with hands on, inquiry based, grade-appropriate science activities to use as part of their pre-service training. If later mentor-teachers do not practice such teaching strategies, pre-service HOL teachers will already have the HOL experiences to draw from and will be able to compare different teaching strategies. This chapter describes the development of geoscience based laboratory activities designed for undergraduate, early teaching experiences in the Hands on Laboratory course at CSU Chico.

The Hands on Laboratory experience was originally conceived over 20 years ago to be an isolated field trip to the university campus for elementary school classroom teachers and their students. Chemistry faculty and science majors conducted the labs several times each semester and the program became known to local teachers who would bring their students to campus for the occasional science experience. It became clear that the HOL could be more broadly used to help train students with an interest in teaching and that it could be an effective component of the classroom curriculum and provide a service-learning opportunity for the pre-service teachers serving as HOL interns.

In 2002 funding for the HOL program was obtained in part through NASA NOVA's granting program (discontinued) for developing integrated science problem based courses, with matching funding from CSU, Chico and from the California Science Project's Teacher Preparation Pathway one time competitive supplement. In 2005 the Internship in Science Teaching (NSCI 489 Natural Sciences) course was adopted as a requirement for all Liberal Studies majors, the major of choice for most of the future elementary school teachers at CSU, Chico.

Goals of the HOL: The primary goal of this classroom experience is to enable future elementary school teachers to envision themselves teaching science. Pre-service teacher training programs in California such as the CSU Chico Liberal Studies major program must provide high quality training in science; both in breadth and depth so that future science teachers are well prepared in content and teaching strategies that provide high quality instruction to their future students. As a service-learning opportunity, the HOL also provides local teachers with the opportunity to bring their students to the lab for instruction in standards-based hands on activities taught by well trained interns.

The HOL Internship in Science Teaching Course-The HOL is one of the most important components of the CSU Chico Liberal Studies program. Students intending to teach at the elementary school grade levels are required to take two courses in Life Sciences, one course in Physical Science, and one course in Earth Science. As they complete their content-knowledge coursework, undergraduate students are also required to enroll in the HOL's Internship in Science Teaching course.

During the first three weeks of the semester the undergraduate HOL interns are introduced to the content they will be teaching to elementary students along with teaching strategies and classroom management skills by experienced faculty and the teacher in residence. Labs are taught to the visiting elementary aged students by the undergraduate interns enrolled in the HOL course, but are constructed by science and HOL faculty, providing interns and their students with activities designed to explore science guided by the California Science Standards in an interactive, engaging manner. Classes of elementary students come to the HOL starting in the fourth week, continuing through the 15th week of the semester providing undergraduates a total of twelve teaching experiences in six content areas.

Elementary students are introduced to the theme of the lab, and then spend approximately 15 minutes on each activity. Labs are 90 minutes long during which the interns guide young students through a rotation of six stations to explore a topic within the main lab theme of the day. Each content area is taught for two consecutive weeks, giving interns the opportunity to improve their pedagogical content knowledge (PCK) through iterative teaching experiences. Repetition of the lesson makes it possible for interns to reflect on the first teaching experience for a specific topic and then discuss, revise, and re-teach the same topic. Class meetings in addition to the teaching experiences include "debriefing" sessions in which interns meet with a faculty instructor to discuss successes and areas needing improvement for lab activities and teaching practice. Interns are also required to keep teaching journals to record their ideas about each teaching experience, reflect on their progress in teaching and content mastery, as well as respond to journal prompts.

Undergraduates become accustomed to reflective teaching in a supportive group setting in the HOL. Rather than accept the lesson as a scripted procedure with only one correct delivery, students adopt a culture of lesson reflection and revision with the faculty instructor based on the goals and the target authence. This approach serves as a brief introduction to Lesson Study, a process of lesson analysis used in Japan and elsewhere and studied by Catherine Lewis as an effective form of professional development for practicing teachers.

Geosciences in the Hands on Lab

Geoscience thematic labs in the HOL include the Rocks and Minerals activities described here, as well as Plate Tectonics, and Water. Additional HOL themes include topics in Physical Sciences (Electricity, Matter, and Magnetism) and Life Sciences (Digestion and Photosynthesis). The HOL Internship course is generally taken at the sophomore or junior level, once students have satisfied the pre-requisite introductory biology course and an introductory physical sciences course in which they are introduced to basic principles of physics and chemistry. These prerequisites do not include a course in Geosciences (although students are required to take a course in Geosciences prior to graduation) so for some students, their first experiences with rocks and minerals and plate tectonics are in the first few weeks in the HOL preceding their early teaching experiences as interns.

Rocks and Minerals Hands on Lab Activities - The Standardised Testing and Reporting (STAR) program in California includes Earth Sciences for both 4th and 5th grades. The science exam is taken in 5th grade and includes both 4th and 5th grade content areas (Rocks and Minerals and Water, respectively). Exam results factor in to each school's Academic Performance Index (API), making the geosciences an important component of elementary school curriculum.

The HOL was designed specifically to provide interns training in the delivery of science content using standards based concepts and unique sets of materials that are often unavailable to individual teachers. Thus, in the spirit of service learning, the need for a set of HOL activities covering rocks and minerals was identified to provide the undergraduate student interns with teaching experience in fundamental concepts in geosciences required by state standards, as well as a foundation for teaching these topics with hands on techniques and reform teaching practices. These activities are also an important resource to local teachers who commonly do not have the expertise and materials needed to teach Rocks and Minerals in their own classrooms.

Design Philosophy of Activities: The Rocks and Minerals laboratory activities were designed to meet the academic content standards for science adopted by the California State Board of Education (2008b) for fourth grade students. The standards require students to know now to differentiate between igneous, sedimentary, and metamorphic rocks, identify common rock forming minerals, and understand that properties of rocks and minerals reflect the processes that formed them.

The thematic goal of the Rocks & Minerals lab introduces rock samples to students within the context of California-specific geology. Six activities were developed with this theme in mind, but because of the rotation format of the HOL, students start at any of the six stations. Therefore each station must be "stand-alone" and none can assume knowledge gained from a prior activity. These stations serve as a skeleton/ augmentation/catalyst for what occurs in the classroom and are not meant to be the only exposure elementary students have to this content.

With our goals and constraints in mind, we developed the following six stations (presented in random order, as the interns and young students might encounter them):

1. Minerals in Rocks of the Sierra Nevada Mountains: interns teach students the differences between rocks and minerals, as well as diagnostic properties of the rock forming minerals that make up rocks of the Sierra Nevada Mountains.
2. Extrusive Igneous Rock Textures: interns lead students in making connections between cooling rates and textures of igneous rocks. Included in this station is a demonstration of the textures formed with rapid and slow cooling of salol. Rocks examined at this station come from local volcanic centres including the Lassen Volcanic Centre and Mt Shasta.
3. Sedimentary Rocks and Fossils: students use grain size and textures to compare sands and river gravels with local sandstone and conglomerate samples. Students also examine microscopic fossils in sand collected locally.
4. Metamorphic Rocks: students distinguish foliated and non-foliated samples of local basement rocks and the Coast Range, including description and identification of serpentinite, the state rock of California.
5. Erosion and Landform Development: interns lead students through a guided inquiry activity in which they explore factors related to stream erosion, coastal landforms, and geologic hazards.
6. Panning for "Gold": students pan for pyrite and galena to explore ideas of density and make connections to topics learned in local history related to the California Gold Rush.

Prior to beginning work at each of the stations, students are introduced to the theme of the activity by the faculty member or lead intern in the lab. Logistical instructions are given, along with information regarding the themes to be explored. The Rocks and Minerals lab theme begins with an introduction to the rock cycle. This orients the students and reminds them of the prevailing theme of the lab activity. Once rotation begins, interns introduce the specific lab activity by introducing concepts within the context of the rock cycle. Interns then guide students through the activity, helping students make connections to previous activities. Before leaving each station, students respond to guided inquiries in their "lab notebooks" to record their observations and conclusions made at each station. Early in the semester, interns typically prefer to teach only 1 station as the elementary students rotate between the stations. Journal entries at the end of the semester reveal that interns have gained enough confidence in the material and their teaching ability to want to teach additional stations and rotate with the students.

Importance of the Hands on Lab to Future Teachers

Since 2004 the Coordinator of the Hands on Lab, Tanya Heaston, a California-certified teacher, directed the logistics of the HOL as she conducted her Master's in Education research on the impacts of the HOL on Liberal Studies majors working in the lab. This research and responses to journal prompts by interns help gauge the experience undergraduate interns have had in the HOL.

Teaching Impacts-The HOL provides pre-service teachers the unique combination of an environment to learn content knowledge and application of that knowledge in a teaching activity. These experiences all take place in the safe environment prepared and monitored by faculty teaching and content experts. The combined content-pedagogy model of the HOL preparation of pre-service students is reported to be highly valued by undergraduate instructors in the HOL. Intern Anna (pseudonym) writes, the HOL "has definitely made science more relevant for me! Before I really didn't enjoy science that much now that Ym teaching it, G m relearning everything in a new way that's more fun. It'll also help me when I become a teacher! I'll feel more confident teaching science now." Another journal prompt asking interns how the HOL made learning science relevant provoked this response from Chanda (pseudonym): "it is completely different learning something for yourself and learning something to teach young minds for me, teaching the science as well as learning it are relevant."

In addition to teaching experience, the HOL provides an opportunity for pre-service students to understand first-hand the importance of mastering topics in science early in their content coursework. This helps students recognise the need for content-based courses they take concurrently or subsequent to the HOL. Intern Ally's (pseudonym) journal includes a note, "The lab was really fun to teach because I really understood the material. Having GEOS 342 (the earth science course taken by Liberal Studies majors) with HOL is useful." As such, the HOL serves as a "just in time" experience in giving undergraduate students context for the importance of their science coursework, widely recognised as improving students attitudes toward science.

Content Impacts-Content knowledge improvements in HOL interns before and after their experiences teaching in the HOL are quantified with pre and post-tests. Test scores showed an increase of 7-71% (42% mean value) for students enrolled in 2005. Sophomore interns (with 31-60 units of college credit) showed the greatest improvement, despite their lack of experience in prior science courses. This suggests the HOL environment is an effective means to learn science for the college undergraduate interns. Sandy (pseudonym) was an intern in the Fall 2006 class and mid-semester wrote a note to her instructor saying, "I have learned more things in this class than I have in all of my other science classes." Similarly, intern "Maddie's" (pseudonym) journal includes an entry, "The HOL has made learning science relevant to me the labs make the science information applicable and understandable." In some cases, interns have already been introduced to content areas in previous courses, but appreciate the opportunity to relearn or review material m the context of teaching it to students. Intern "Gwen" (pseudonym) included a journal entry stating that the HOL "really helps me to understand the labs and their concepts when I teach it [sic] myself... they seem to teach and reiterate difficult concepts that I have already learned." This substantiates the success of the combined content-pedagogy model for preparing undergraduates to teach.

Attitudinal Impacts-Using pre- and post- attitude surveys Heaston (2006) found that after their experience in the HOL, undergraduate students reported improved attitudes towards teaching and expressed greater confidence in their content knowledge of material covered in the HOL. Intern "Kathy (pseudonym) expressed this in a journal entry, " the class also helped to make me less scared to teach science." This is similarly expressed by "Gina" (pseudonym), "the most beneficial part of the HOL lab [sic] for me was getting rid of my phobia for teaching science. It has helped me to gain confidence in myself as a science teacher. I love that I can stand in front of a group of kids and teach them without worrying." A journal entry from intern Genevieve (pseudonym) includes, "The HOL has been amazing. Coming into this class I was not at all excited about science because I didn't understand a lot of it. Going out I'm taking a lot with me."

Attitudinal surveys analysed by Heaston (2006) indicate that sophomores, juniors and seniors all report increased confidence in their being "prepared to teach science at a variety of grade levels" (mean scores increased by 0.5-1.03 points out of 5 points, or 5-20% improved attitude ranking), interns also report improvement in their confidence in being able to "turn students on to science," their confidence in their ability to teach science, their interest and confidence in the use of technology, and knowing that it will be useful to study science in their preparations to teach. Intern "Wanda" recorded in her journal that, " the HOL has given me a great overall and in-depth perspective on science really enjoyed the HOL class because the teaching techniques and opportunities really gave each intern a chance to work on teaching skills and feeling comfortable and in control in front of a group of students."

The HOL has played an important role in improving the interest and confidence of pre-service teachers in teaching geosciences in their classrooms. Student (intern) attitudes towards learning science in the HOL are represented by an additional entry by "Kathy" (pseudonym), who summarises the value of the HOL experience in her entry, "I was very apprehensive at the beginning of the semester about teaching science, and this class has helped get rid of those fears all our labs provide great ways of teaching different concepts." Intern Kaitlin Banks reported to the student newspaper that she, "likes the class because it gives her first-hand teaching experience, and the lesson plans make the material easier to understand and teach: she gets nervous every time she teaches a lab, but she knows the experience will help".

The Hol Beyond Teaching Preparation

Graduates of the HOL intern course generally graduate with a Liberal Studies degree, complete a credential program and begin teaching. The impact of the HOL remains with them as teachers, both in the use of HOL teaching methods, as well as in their continued value of science. "Sarah" (pseudonym) graduated from the HOL program in 2007 and now teaches 2nd grade and recently wrote her former HOL instructor to say that the HOL, "really prepared me for teaching science to children. If it hadn't been for the class I don't think I would have felt as comfortable or prepared as I did."

Teaching Materials-In addition to improved confidence levels, Liberal Studies majors also benefit from the HOL in that their experiences in the lab directly translate to a portfolio of teaching that students can use to document their experiences and skills for future use. Currently a third grade teacher, "Addy" (pseudonym) corresponded with er HOL instructor while in her credential program saying, "I am actually using the (HOL) binder with all the labs in it my credential program university does not have a class like the HOL at CSUC, so I'm probably the only one in my [credential] class that has taught science before!!" Another HOL graduate recently wrote to his former instructor that he uses his HOL course materials in his teaching now. "James" (pseudonym) recently wrote, "I love to be able to call upon all of the experiments we did, I still have the write-ups from the labs I did."

HOL Resources for In-service Teachers-The HOL has become a highlight of summer workshops for teacher in-service professional development program for several years. The California Science Project of Inland Northern California (CSP-INC) offers summer institutes in which teachers learn about a variety of teaching and learning strategies in science, generally linked to a common theme. Experienced interns and workshop leaders guide in-service teachers through several HOL themes so they can a) learn content in a hands on environment and b) implement similar activities in their own classrooms. To support their efforts workshop participants are offered priority registration to bring their classes to visit the HOL during the academic year. Teachers attending the CSP workshops rank their experiences in the HOL the highest of any professional development experience offered. The HOL experiences typically rank 4.8 on a scale of 1-5 and result in self reported increases in interest, comprehension, and ability to teach topics in geosciences in their classrooms.

The HOL has also developed science kits with materials that teachers can borrow for use in their classrooms, covering topics offered in the HOL. All teachers (intern graduates and in-service teachers) have access to loan materials available from the HOL. One indication that the Rocks & Minerals activities in the HOL have been useful to teachers is that during the developmental stages of the kit library, 3 of the 8 kits loaned during the Fall 2007 semester were for Rocks and Minerals. Teachers who use the kits typically find them easy to adapt to their classes and teacher "Bernardo" (pseudonym) reports that his students were "so busy having fun, they didn't even know they were learning concepts they will use later in school and life, (and) in my third grade class, my students didn't want to leave the class to go to clubs. They didn't want to leave because they were having so much fun."

The impacts of the HOL though, have broader reach than the Liberal Studies majors, and teachers; the young students visiting the HOL also receive a high quality science experience. "If really makes science come alive and really gives it meaning to the students" says fifth grade teacher Kathie McKenzie of the Rocks and Minerals labs.

Positive feedback in undergraduate student evaluations and journals along with the encouraging results of Heaston's (2006) work demonstrate the success of the HOL at improving undergraduate pre-service teachers' attitudes in teaching science, including geosciences. Increased content knowledge is similarly achieved as interns learn material they need to teach to their students. These results and similar endorsements from all levels of HOL participants (students, interns, teachers, university faculty and HOL staff), inspire our commitment to the HOL and further motivate pursuit of our goal to increase content modules for additional lab themes in the HOL as well as to help in development of similar programs at other colleges, universities and community colleges across California and the U.S.

Improvements to the HOL are ongoing. Continued complexities include maintaining high quality classroom materials; managing logistics with teachers who bring their students to CSU Chico's campus to experience the HOL but are faced with budget crises that restrict funding for transportation.

Cohesive training for interns is an ongoing challenge, as is merging undergraduate intern's schedules with those of the visiting classes. New plans to expand activities of the HOL include introduction of new lab themes. A new series of labs currently in development address the theme of Energy, a complicated but important subject for future science teachers in general, but specifically for future geoscience teachers.

Longer term plans include broadening the grade levels served by the HOL to K-3 (pending funding) and development of activities for high school level content, thereby broadening the grade levels that interns teach. We look forward to continued work in the HOL to encourage undergraduate interns, teachers, and pre-college aged students visiting the HOL to continue to have positive experiences in geoscience and all sciences that with the goal of impacting younger generations of scientists and educators.

Changing Priorities in Geoscience and Science Teaching

Science education seems to be in constant crisis, at least in UK schools, as indicated by student enrollments in post-16 science courses (it is mandatory to 16 years) and students' declared interests, priorities and career objectives. Potential remedies lie both within and beyond science education and include the focus of this chapter: argumentation pedagogy. In contrast, school subjects within the social sciences and humanities are thriving: notably psychology, philosophy, economics, history, citizenship and religious education. All are subjects which routinely involve debate, controversy and argumentation. The perception of science as a cold, controversial, divisive and potentially anti-social activity persists among students, summarised by Ramsden (1998) as: "the widely held perception of science being difficult and not relevant to the lives of most people; of science causing social and environmental problems; that science is more attractive to males than females; that interest in science decreases over the years of secondary schooling; that these more negative views are associated with the physical sciences rather than the biological sciences".

Given the unsatisfactory position of science in schools, including geoscience, what steps can be taken to improve matters? Many: we are not short of answers to this question. Several of them involve the use of argumentation in teaching, a widely researched approach for fostering learning among school students and their teachers. In recent years this has become a growth area in science education research, with an increasing number of studies across the sciences. Here I examine some specific issues that arise in relation to geoscience and argumentation.

The UK curriculum changes introduced from September 2006 represent a distinct loosening of control by government, with schools being given wider choice of curriculum content and breadth, although science remains a compulsory subject for all students to the age of 16 years. Current developments also include (i) the systematic inclusion of thinking skills, (ii) a higher profile for science in society (SIS) issues, notably linked to the mandatory subject of Citizenship and (iii) increased attention to developing scientific literacy, a label first used systematically in the 1950s and now used increasingly in relation to the aims of new science curricula on a global scale.

Built on a major research-based report, "Twenty First Century Science" is an innovative and very flexible UK science curriculum for

15 and 16-year-old students designed explicitly to enhance scientific literacy. Structured explicitly around biology, chemistry and physics, it provides a 2year course for all students, including those likely to study science to an advanced level ("science for all and science for scientists"). In relation to curriculum content, the authors of the programme decided that "what citizens require is a broad, qualitative grasp of the major science explanations; the detail which many students find offputting is rarely needed". Such content was expressed as sixteen "core explanations", several of which are predominantly geoscientific (eg "The structure and evolution of the Earth" and "The theory of evolution by natural selection") and most of which have geoscience connections (eg "Energy sources and use" and "Materials and their properties"). More important for geosciences educationalists than any detailed content is the rationale offered for the composition of this list: "the primary selection criterion [for content] was that an explanation should be included only if an understanding of it might make a difference to a decision or choice that a citizen could have to make, or to the viewpoint he/she might hold on an issue or decision at local or national level, or if it offered a culturally significant view on the human condition (on our ideas about 'who we are and where we are'). The latter, for example, justifies the inclusion of ideas about human evolution and cosmology which have less obvious practical implications". Such a rationale has immediate implications for teachers since such aims are unlikely to be met through traditional didactic approaches: a participatory pedagogy is required. Argumentation meets this requirement.

Science is a sociocultural activity: people engage in scientific endeavour collectively and that collective activity contributes towards that nation's cultural identity. Any nation's schools have a duty to develop scientific literacy among its pupils so that they can participate in democratic debate on scientific matters of significance. Citizens also need the skills to discuss the nature and purpose of science, skills that can be developed in school. Argumentation is the antidote to archaic perceptions of science, which still persist in some UK schools, as a tightly positivist, fact-ridden, unimaginative and formulaic set of procedures, largely divorced from emotion and certainly moving away from the errors of past science activity. Argumentation accepts that people engage in dialogue as part of their science work in order to advance knowledge and understanding. It represents a powerful tool for developing children's science learning, especially their higher order cognitive skills of understanding, application, analysis, synthesis and evaluation. In this chapter I suggest several ways in which argumentation can be used in the context of geoscience teaching and learning.

Argumentation, as I am treating it here, is a form (or genre) of human discourse which: (i) is rational; (ii) has several distinctive components; (iii) involves both intellect and emotion (affect), and; (iv) is designed to resolve or advance an initial claim or opinion through persuasion based on evidence. The term 'argumentation' refers to the process whereas the term 'argument' refers to the initial proposition and its subsequent debate: the 'content' around which the process of argumentation takes place. Mason and Scirica (2006) provide a pithy and well referenced account of argumentation skills in the context of classroom-based teaching and learning. 'Argumentation' and 'debate' are not quite synonymous since the first implies the systematic implementation of agreed procedures (and labels), whereas 'debate' has a great diversity of meanings and connotations because it refers to any form of structured discourse.

It is only in the last few decades that argumentation has entered science classrooms in a structured and formal way, notably as part of the desire to focus on children's scientific literacy and wider thinking skills. It is part of the shift towards a more explicit recognition that scientific knowledge is socially constructed by learners and that structured discourse leads to enhance conceptual learning. Indeed, Kuhn (1993) argues that it is the argumentation process which most closely corresponds to the thought processes in science, claiming that "we can find scientific thinking in older children, adolescents, and lay adults if we conceive of it as argument". She concludes that "science as exploration and science as argument do not in fact contradict one another" and that "the natural curiosity that infants and children show about the world around them needs to be enriched and directed by the tools of scientific thought". This constitutes the underpinning rationale for the current chapter.

Argumentation also has the potential to improve students' attitudes towards science, their enjoyment of science learning and their powers of reasoning. Thorough descriptions and analyses of the role argumentation in science and in science education are given elsewhere. The following is a brief summary of these ideas, with some illustration with geoscience concepts.

Stephen Toulmin's classic book marked the start of the modern interest in argumentation as a process of reasoning between two or more people which incorporates both cognitive and affective skills. He had a special focus on legal discourse. Since then, other authors have presented his approach as a model comprising four core elements, labelled 'data', 'claim', 'warrant' and 'backing' and many authors have refined the model beyond these elements, especially researchers seeking to develop and evaluate new ways of implementing an argumentation pedagogy. As the opening stage in an argumentation process, a 'claim' (proposition) is made by reference to 'data' (information). This might be something such as 'the continents of Africa and South America were once united [claim] because identical fossils and rock types can be found on matching coasts and hinterlands data'.

A third element comprises a 'warrant', which gives the link between the data and the claim: the reason WHY the claim arises from the data. In this example two relevant warrants might be Organisms that can only survive in shallow seas cannot cross the deep Atlantic Ocean' and 'identical fossil species are almost always of the same geological age'. A fourth key element is called the 'backing', which comprises the agreed basic assumptions in the system: here they would include organic evolution, preservation of fossils and the formation of sedimentary rocks, but they would not include plate tectonics (or drifting continents) since that forms the central claim. In addition to the four central elements identified above, we can recognise 'rebuttals' and 'qualifiers'. A rebuttal is a challenge to the claim (a counter-claim), with supporting evidence. In other words, a counter-position is taken up and appropriate evidence is cited. A rebuttal in our continents example would be the claim that landbridges and shallow seaways existed between Africa and South America and that the continents have not moved in relation to each other. Evidence would include the various mid-Atlantic islands such as Canaries, Cape Verde, Falklands, Tristan da Cunha and St Helena. A 'qualifier' is a refinement of the original claim: a statement of the conditions under which the claim is true. In our drifting continents example, a qualifier might refer to variation in possible speeds of continent separation (eg between 2 and 5 cm. pa) and, therefore, possible dates for initial rifting.

A few more concepts (and their name labels) can help in the analysis of argumentation. When claims (or propositions) are made, it is important for children to be able to distinguish between the following basic scientific concepts: 'facts' (universally agreed propositions), 'assertions', 'hypotheses' and 'theories'. During argumentation, children draw on 'evidence' to support a claim, rebuttal or a qualifier: this evidence may be 'personal' or 'authoritative'. They also adopt strategies which may be designed to cause or accentuate either 'conflict' or 'collaboration', which in turn might lead either to an agreement to differ ('coexistence') or to a 'consensus'. Strategies may be cognitive (eg elaborating or clarifying an explanation) or they may be social (eg body language to signify disapproval). Furthermore, they might seek allegiances with others to enrich their claim or rebuttal through cognitive or social means; or they might seek to act alone.

Finally, some authors have recognised 4 stages in the argumentation process, at least in the context of science experimentation: (i) an initial 'focus' when the ground is prepared; (ii) a second 'exchange' period when ideas are discussed in order to agree the way forward; (iii) a third 'debating' stage when the claims are actually discussed, and; (iv) a final 'closing' stage, which may be explicit, implicit or circumstantial, reflecting declining levels of final agreement.

The Roles of Interest, Motivation and Prior Knowledge

If argumentation is to be used in the classroom to foster geoscience learning, the inter-related issues of interest and prior knowledge and understanding must be addressed carefully, for both students and teachers. First, in relation to students' geoscience interests, it as been shown that academic attainment is more closely related to 'individual interest' (the robust, stable, long-lasting form of interest, developed over some time) than to 'situational interest' (the transient, temporary, short-lived for of interest typically generated by teachers in classrooms). Second, in relation to students' prior knowledge, there is nothing worse for classroom discourse than ill-informed dialogue that degenerates into vague generalisation, prejudiced assertion, personal attack and all-round disaffection. In our 'drifting continents' example, students' prior knowledge and understanding of plate tectonics and palaeogeographical principles and theories would seem to be pre-requisites. However, some research shows no correlation between the quality of students' argumentation and their prior knowledge of the topic in question, beyond a basic level. Although it appears to be counter-intuitive, "we cannot assume that increases in content knowledge necessarily improve the quality of argumentation".

It seems likely that a threshold effect is present: a minimum level of knowledge and understanding of the content is necessary, but beyond that there is no relationship between students' prior knowledge and their argumentation quality.

This assertion raises many research questions which cannot be addressed here, such as: what is the minimum level of knowledge required for effective argumentation; how can teachers determine that minimum level for any given curricular context; how does this basic geoscience knowledge level vary between students (do some need to know more than others to participate well?); which facets of geosciences cognition are enhanced through argumentation (knowledge, comprehension, evaluation, analysis, synthesis)? The problem of the required knowledge-content threshold for effective argumentation is critically important yet impossible to address systematically since it varies for each student and the topic being studied. The solution lies with the craft skills of the teacher: knowing how much contextual knowledge is essential for each stage in the argumentation procedure and for each student.

When it comes to implementing an argumentation pedagogy in the geosciences classroom, issues of motivation, confidence and geoscience subject knowledge and interest relate to teachers as much as to their students. In particular, teachers need to be supported in the school workplace through the provision of the three basic psychological needs of autonomy, competence and social relatedness. Like their students, teachers are learners who flourish when these three ingrethents have their impact, so school managers and leaders, fellow teachers, education academics and others have a role to play in supporting teachers as they develop new pedagogies.

First, briefly, the need for teacher autonomy implies a professional freedom to make informed judgements concerning their students' learning needs. An argumentation approach is unlikely to succeed if the teacher is a reluctant participant. Second, teachers must have the resources (time, equipment, accommodation) and possess the professional knowledge and understanding which enables them to bring argumentation to their classroom.

Feelings of competence are insufficient: teachers must be competent. Finally, the pedagogical approach needs to stimulate the teacher's feelings of relevance and relatedness to the wider social context, notably the body of teachers and educators within and beyond the school. These comprise the source of each teacher's sense of professional purpose. The acknowledgement and addressing of these three categories of psychological need will narrow the gap between esoteric rhetoric and classroom practice. This issue is further addressed below in the context of a three level hierarchy of epistemological understanding.

The Role of Students' and Teachers' Epistemological Understanding

Several factors are known to inhibit argumentation, including: well-developed personal belief systems; a level of topic knowledge and understanding which is below some (usually unspecified) threshold; a psychological disposition towards maintaining warm personal relationships; a lack of willingness to learn from argumentation; lack of opportunities provided by the teacher, and; a low level of epistemological understanding. On this last point, recent empirical research into the ontogenetic changes in epistemological understanding indicates a progression through several stages from an absolutist to an evaluativist approach, although there are many variants of this scheme. This work builds on the pioneer writings of William Perry (1970), based largely on ontogenetic cognitive development among university students. Hofer and Pintrich (1997) provide a substantial critical review of these research programs. Recent empirical work shows that the learning process itself evolves with human growth, notably in relation to epistemological understanding, but such changes vary enormously between individuals and some results "suggest that some 12-year-olds have become as capable as many adults in managing the interaction of theory and evidence in their own thinking, in a way that supports effective learning". This range of competences among students clearly has direct implications for teachers' management of classroom activities on argumentation.

In their empirical study involving 62 children aged 13 years, Mason and Scirica (2006) relate a three-level hierarchy of epistemological understanding to argumentation pedagogy in the context of global warming and genetically-modified food. Their hierarchy comprises the categories of: (i) absolutist, (ii) multiplist, and (iii) evaluativist. Most children function within the first stage, so they "believe that knowledge is absolute, certain, nonproblematic, right and wrong, and does not need to be justified since observations of reality or authorities are its sources". Such a stance clearly has serious implications for geoscience educators and their students' perceptions of authority, not least in relation to matters of deep time, evolution and young Earth creatsationm. Compare, for example, the learning potential of a student who accepts the age of the Earth (or evolutionary change) as given by (any) authority and that of a student who adopts a questioning stance towards such assertions.

By adolescence, most people's epistemological understanding has advanced to the "multiplist" level, so they tend to "believe that knowledge is ambiguous, idiosyncratic, and each individual has his or her own views and truths". Many adults (including perhaps most teachers) attain the "evaluativist" level, when they "have an epistemology grounded on the belief that there are shared norms of inquiry and knowing. Thus, some positions are reasonably more justified and sustainable than others". Although it has been suggested that "adults of all backgrounds are highly likely to make the transition from absolutism to the multiplist acceptance that knowledge is uncertain and divergent claims legitimate", the same authors report that "no more than half of adults of any background and in any judgment domain make the subsequent transition to the evaluativist position". This should give us cause for concern because it suggests that some teachers possess a more sophisticated epistemology than others, permitting them to guide their students in more powerful and effective ways as they engage in argumentation.

Oral communication is at the heart of argumentation, so it is important for teachers to interpret student utterances. Some students' expressed thoughts may reflect their stable and secure understanding of the concept or phenomenon whereas other expressed thoughts may be transient, unstable, and tentative for which the student has no commitment. In other words, teachers need to understand their students' communications as they use them to diagnose learning. Of course, in most cases a progressive refinement of student understanding occurs alongside the child's development of oral language and growing confidence.

The level of epistemological understanding has several implications for teachers when they devise learning activities involving argumentation. First, classroom argumentation involving young children is likely to focus on agreed knowledge derived from authority: any attempts to challenge such authorities may cloud the discourse and inhibit argumentation. In other words, there needs to be acceptance that the child functions at the 'absolutist' level. Second, students functioning on a 'multiplist' level (ie middle-level) are more likely to accept the existence of different viewpoints and to recognise contradictory interpretations of evidence than are the 'absolutists':

this can enhance the quality of argumentation. Third, although the three levels have been cited as age-related, some regard them more as dispositions than intellectual competences. This means that teachers are likely to find both 'absolutists' and 'multiplists' in their class, and perhaps even some 'evaluativists' who will respond most favourably to any outcome of argumentation!

Geoscience and Argumentation

Geoscience lends itself to argumentation approaches to pedagogy for several reasons. First, many geoscience topics can be labelled as 'socioscientific' which lend themselves to learning through argumentation. Global warming and natural hazards are good examples. Second, geoscience materials are readily accessible, both physically and intellectually: rocks, fossils, oceans, landscapes, minerals and so forth. Third, much geoscience operates on scales which are readily understood by children, typically measured from centimetres to hundreds of kilometres, rather than the micro-scales of nano-science and the macroscales of astronomy. Indeed, it has been reported that "the main indicator of whether or not a high quality of argument is likely to be attained is students' familiarity and understanding of the content of the task". Fourth, geoscience issues are often reported by mass media, so children are familiar with many of the concepts or, at least, their name labels.

Although there have been many reports and theoretical discussions on the development of thinking skills in geoscience classrooms, little research has been published about the systematic use of argumentation in geoscience teaching and learning. There are of notable exceptions and many of the argumentation examples used in the wider science education literature include familiar geoscientific phenomena such as global warming. Geoscience also figures in studies of students' writing skills. Here I want to examine ways in which students can engage in structured and systematic argumentation around judiciously identified geoscience topics, with a further benefit being a more secure grasp of that epitome of a geosciences threshold concept: deep time. To that end, one example is developed in some detail (coastal processes) and five further examples are summarised in tabular form.

In the following discussion the educational emphasis is on the science itself, in contrast to the oft-stated perception that it is the socioeconomic implications and consequences of the science that best lend themselves to an argumentation approach. As a result of this widelyheld view, much educational research and practice in this field tends to emphasise the socioeconomic aspects of science because these are perceived as lending themselves to the development of contrasting opinions more readily than is the science itself. For some teachers this is true: they find it easier to devise learning activities which bring out the political, socioeconomic and cultural aspects of science, rather than the science 'per se'.

This view has to be challenged since science is often a social activity, involving argument derived from scientific evidence. The contrasting positions adopted by scientists are not simply 'opinions' in the colloquial sense, based on their world views or belief systems; they are positions which arise from their professional judgements made after considering the evidence (albeit influenced by their values, attitudes and belief systems). The problem for teachers, however, comes when the children have 'absolutist' views of science: they are unable (or find it difficult) to engage with opposing viewpoints and see no need to justify any claims. Perhaps argumentation provides an opportunity for teachers to develop pupils' skills in metacognition so they move closer to a 'multiplist' position. Accordingly, in the following discussion, a central aim is to identify contrasting scientific viewpoints, rather than opinions, which can be defended or challenged with reference to evidence obtained from the world external to the student.

Coastal Processes Around Bays and Headlands

This detailed exemplification of argumentation is based on the idea that many coastal features, notably bays and intervening headlands, can only be understood with reference to sea level change through deep time: two key variables that are typically omitted from UK school texts and websites. Children often encounter the idea that coastal bays are formed by the marine wave erosion of 'soft rocks' and that headlands result from 'tough rocks' which resist such wave erosion. Bays arise directly from marine erosion. However, it is easy to demonstrate to students that extant processes operating in such environments usually show maximum erosion on the headlands, in a high energy environment, and maximum deposition within the bay, typically marked by a large bay -head beach. In other words, there appears to be a contradiction between commonly held ideas and the field evidence. Such a contradiction is easily presented to students. And it is this property which renders this topic ideal for an argumentation pedagogy. It is particularly effective when the various ideas are introduced to students in the field on a stormy day! The apparent contradiction provides the stimulus and starting point for the argumentation process.

One possible approach is to foster children's argumentation using the 'competing theories' approach. First the apparent contradiction is presented, as above, to stimulate thought and informal discussion, ideally before, during and after a field visit. According to student ages and abilities, the teacher then stimulates hypotheses or prompts students to suggest possible theories to explain the phenomena. Claims can then be offered at this initial stage, with supporting evidence, again ideally in the field. These opening claims might lead to counter-claims and thence continued dialogue (debate) back in the classroom. In short, the competing theories are generated by the students by drawing on their previous learning and, inevitably, much depends on the management of this classroom discussion.

Alternatively, several possible theories (three in this case) can be presented to students directly as resource cards. Students can be arranged into small groups or pairs to examine and defend each theory. Whatever the source of the theories (student or teacher or both), they can be analysed and evaluated as appropriate. The text should be analysed and, for example, facts, assertions and theories should be distinguished from each other. Once the several competing theories have been studied and perhaps summarised, the students need to engage with potential evidence which may be obtained from field activities or it may be presented by the teacher, or a combination of both. The teacher may introduce items of evidence as another set of cards. Each of these can be introduced at strategic times during the discussion: a drip-feed approach. Such evidence cards would then have to be sorted by students as either (i) 'potential evidence in favour of a theory', (ii) 'potential evidence against a theory' or (iii) 'potentially irrelevant'. In addition to the evidence cards, or possibly replacing them, further evidence can be called upon by students to develop their case, typically from the Internet or from field observations.

Further resources can be introduced to enrich the debate and provide for qualifiers to be identified. For example, a map sequence to show the geomorphological evolution of the relevant stretch of coastline over time can be presented, with the added ingrethent of a sorting task to place the maps in their correct sequence.

The range of epistemological understanding among students impacts on the nature of this coastline discourse, but not necessarily on its quality, and one of the key purposes of this pedagogy is to enhance such student understanding. The teacher should identify students starting from an 'absolutist' position and take care to guide them towards a consideration of evidence which suggests alternative theories, thereby assisting their passage to a multiplist position. Similarly, 'multiplist' students should be encouraged to evaluate the range of theories, perhaps ranking them in order of efficacy (an 'evaluativist' stance).

Other approaches to encourage learning through argumentation over this sea-level change example can be developed. For example, the map sequence mentioned above can form the main item of evidence, with the main challenge being to explain how the present day coastline has evolved. The two most important points in both approaches are that (i) students use structured argumentation to engage in real debate over possible explanations for coastal evolution and (ii) the deep time perspective is added to the equation: it is often ignored. A third issue which can be included relates to the argumentation process itself: students can be taught the techniques, labels, concepts and protocols of argumentation in order to reflect on their own learning and thereby improve the quality of their argumentation skills. Such an approach has been shown to enhance learning.

The extent to which argumentation can result in enhanced geoscience knowledge and understanding depends not only on the quality of the arguments (contents) but on the social interaction between participants; so the classroom climate is critical. Ideally, claims based on observations and deduction are challenged or supported with reference to wider lines of evidence, eventually coming to include the full range of past and present variables known to be impacting on coastlines. In this example, coastline evolution through deep time should be included as the critical explanatory factor, rather than excluded as being irrelevant to the development of bays and headlands. Some geoscience topics lend themselves to an argumentation pedagogy more readily than do others.

Geoscience appears to be in a strong position to further the development of classroom argumentation strategies because it includes so many concepts and processes that have high profiles in political, educational and sociocultural debates. The science behind critical and familiar processes such as climate change, seismic activity, coastal and river flooding and potential asteroid impact needs to be taught in the classroom in ways which stimulate and motivate children.

Argumentation has the potential to meet that imperative. However, interest, motivation and attitudes towards knowledge cannot be ignored. The pivotal issue of epistemological understanding needs careful attention if this approach is to pay learning dividends since each child perceives knowledge, and its source and authority, in their own distinctive way. Teachers need to be aware that such diversity in perception leads to contrasting attitudes towards matters such as knowledge, argument, evidence, theory and proof. The categories of absolutist, multiplist and evaluativist provide teachers with a logical and structured framework in which to address such diversity in the classroom, giving them opportunities to apply their craft skills and their professional knowledge of their own students in developing an effective argumentation pedagogy.

Writing as a Learning Tool in Geoscience Education

As faculty members in geoscience education, we face a quandary with student writing: on the one hand, the quality of their writing is often below expectations-including grammatical sloppiness, poor transitions, underdeveloped or under-supported ideas, or a general inability of our students to write like geoscientists. On the other hand, we are resistant to expand the writing component of our classes because the process of grading writing can seem subjective, inexact, and overly time consuming. Few of us may be enthusiastic about teaching a writing-intensive course or revising one of our courses to become writing intensive. Of course we realise that if we do not provide our students with practice in learning to write like geoscientists, they will nave little opportunity to develop such skills. Offloading this responsibility to a technical writing or composition course loses our disciplinary expertise because our content knowledge puts us in the best position to help students understand how disciplinary thinking informs the writing and content of geoscience documents.

This chapter addresses these issues and provides another way of thinking about incorporating writing in geoscience education. Our authorial perspectives stem from our disciplinary backgrounds; one of us teaches writing-intensive geoscience courses in a geology and geological engineering department, and the other conducts research and faculty workshops on Writing Across the Curriculum (WAC) and Writing in the Disciplines (WID), particularly in engineering and applied science settings. The impetus for this chapter came following one of these faculty workshops, when we recognised that writing in the geosciences could be improved and expanded if we could demonstrate practical, research-based strategies that enhance learning and minimise the labour of preparation and grading. Before demonstrating such strategies, this chapter will provide both a general and a geoscience-specific background on writing as a learning tool.

The notion that writing is a learning tool is framed by two important areas of research, in WID and in writing and cognition. Generally, WAC programs aim at improving students' higher-order thinking and problem solving skills as well as their ability to communicate ideas. Unlike more transitory educational reform movements, WAC in 2005 celebrates its 35 year anniversary in the US higher educational system. Its staying power is attributed to numerous factors, among which is the fact that WAC helped catalyse and develop numerous staples of current higher education pedagogy, including thematically linked cluster courses, learning communities, collaborative learning, and peer tutoring. Also among WAC's most salient attributes is that writing facilitates problem solving and higher-order thinking. While other educational fads have faded from sight, WAC has experienced steady growth. The fledgling WAC efforts that began in 1970 were widespread by 1988, when approximately 50% of all US postsecondary institutions had WAC programs, a percentage that continues to rise.

As a complement to and often functioning within WAC, WID efforts focus on writing within each discipline or subdiscipline and have several research-based premises:

- That writing is not learned in one class at one time but requires continual practice
- That writers establish a broad foundation through general writing courses and then require specific guidance from disciplinary faculty on writing within their discipline
- That when used appropriately, writing involves active learning and student-centred instruction.

WID is also premised on the notions that "writing expectations and performances are discipline specific, that "much of what an accomplished writer knows about writing is tacit knowledge," and that this "tacit knowledge is conveyed largely via apprenticeship". Such premises reinforce the idea that geoscience educators are in the best position to prepare students for their professional writing responsibilities by guiding students' understanding of geoscience writing. This chapter facilitates that process by providing geoscience faculty with a conceptual framework on writing as a learning tool as well as practical ideas for using writing to facilitate learning in geoscience courses.

Writing facilitates learning in numerous ways. Long-standing research on writing and cognition indicates that writing facilitates higher-order thinking and the learning of content in any discipline. Reviews of qualitative WAC and WID studies indicate that although writing is not the appropriate learning tool for students doing traditional fact-based learning, it is well suited to more open-ended problem solving. One key advantage of WID is that "students must pick up not only textual features as they learn to write, but also the ways specialists think...". Indeed, because disciplinary writing is shaped by discipline-specific habits of mind and methods of inquiry, teaching students to write like geoscientists is inherently interrelated to teaching them to think like geoscientists. A review of 35 research studies on writing as a learning tool indicates mixed results, with studies suggesting that in concert with other factors, writing can foster concept synthesis and introduce students to a disciplinary community. More recent research on writing to learn science has accentuated the learning benefits for secondary students. When implemented appropriately, writing has been associated with multiple learning gains, such as making increased conceptual connections, recognising one's own knowledge gaps, distinguishing between claim and evidence, and metacognition. Learning gains have also been noted among university students using writing to learn science, especially when certain instructional practices are in place.

Research on the cognitive processes associated with writing also substantiates the complexity of the task of writing. Writers tax multiple cognitive resources as they shuttle between long-term memory, where they store content, audience, and writing knowledge, and short-term memory, where they store planning, translating, and reviewing processes. This and other research also indicates that effective writers do not possess content knowledge alone, but combine this with knowledge about discourse-about the ways of inquiring, reasoning, and communicating that are common among professionals in a given field.

Despite the evidence linking writing and learning, barriers to successfully incorporating writing remain, and many stem from the normal struggles of students and faculty. Research and our own teaching experiences indicate that students struggle with understanding many aspects of writing assignments, especially understanding new genres (including genre-specific purposes, audience expectations, and more) as well as new content.

As mentioned above, faculty often struggle with the time involved and inexactness associated with grading, and some may think the benefits of assigning writing do not outweigh the associated costs. Yet writing can, instead, serve as a way of addressing both student and faculty struggles and as a tool to facilitate learning of geoscience course content.

4

Investigating Geocognition

The best geologist is the one who has seen the most rocks. What does it mean to be the best, most expert geologist? Is it simply, as the saying goes, the ability to recognise the most rocks? Or, does expertise differ in more subtle or complex ways? The nature of geological thinking is both poorly understood and poorly studied from an empirical perspective, although numerous investigators have discussed what geoscience expertise might look like.

Only recently have researchers begun to systematically investigate how geoscientists engage in authentic practice, in both simulated and real settings. Bond and colleagues (2007) show that prior knowledge and experience play a large role in conceptual uncertainty, essentially biasing expert interpretations of seismic sections. Kastens et al. (2009) discuss the relationships between novice and expert mapping behaviour and interpretation of artificial rock outcrops, finding clear differences between how these two populations see and understand these phenomena. In one of the few studies of expert behaviour in a natural field setting, Gahegan and Brodaric (2001) provide evidence for the influence of situated cognition in map generation. Other forms of geologic observation and data interpretation are probably also laden with interpretive, situational, and experience related distortions. Given the paucity of existing studies in natural or contrived settings, we are forced to make assumptions about the nature of geological expertise, and we still know very little about how novices (undergraduate students) become experts (professionals), what skills are important to gaining expertise, and how traditional education shapes geocognitive development. Geoscientists universally regard field mapping as necessary for the development of expertise in the geosciences, and geologic field mapping courses play a significant role in college and university departmental curriculum and expectations. Anecdotal evidence suggests that this a priori assumption exists across geoscience disciplines, almost regardless of the importance of field work within that discipline. Within any geological sciences discipline, nearly all postsecondary students receive some field-oriented training, typically as a semester course in field methods, and/ or a four to six week summer field course. Despite this infusion into the curriculum, the nature of expert and novice field-based geocognition, as well as the interaction between cognition, behaviour, and environment during field activities, is open for debate.

By utilising data and method triangulation, we propose to establish a methodology for investigation of field mapping behaviour and cognition. In social science research, triangulation refers to a specific mixed methods research design in which qualitative and quantitative data are gathered and analysed concurrently yet independently, so that the results of each data set can be compared to one another. The triangulation of multiple data sets collected by one or more observer, use of competing theoretical schema, or application of multiple methods to analyse collected data are all insightful mechanisms for enhancing the rigor of research findings. This is particularly true in the cognitive or behavioural sciences, where interpretations can be biased towards preconceptions, and where all aspects of validity and reliability can be difficult to assess.

Our approach utilises multiple qualitative and quantitative data sets, with multiple methods applied to investigate our quantitative data. Specifically, we collected and analysed self-reports of prior geology experience, field maps, and navigation data from all participants, and Think-Aloud audio logs gathered during mapping plus a follow-up interview from a subset of participants. We suggest that this mixed methods approach offers richer and more robust understanding of field cognition and behaviour than that available through simpler single data set analysis. We developed and tested this methodology at two field sites in the Michigan Upper Peninsula, USA, using data collected from seven participants; here, we describe the methodology and results of this study, and make recommendations for using these methods to study authentic geocognition in the field.

Expertise as a generic construct has been investigated in a wide variety of disciplinary areas such as chess, history, teaching, and physics, providing some general insights into the nature and development of expertise, as well as into the application of expertise studies to education. Experts display rapid recognition of patterns that are common in their disciplinary area, thinking that revolves around big ideas (rather than memorisation and recall), interrelated knowledge, the ability to self-monitor and evaluate progress on a disciplinary task (metacognition), flexible thought processes, and awareness of complexities and abstractions.

This body of research suggests that explicitly attending to aspects of expertise, including how experts solve problems, can align student thinking closer to expert practice.

For example, research in physics education shows that experts differ from novices in having knowledge that is hierarchically organised around key principles, and a highly metacognitive approach to problem-solving using such principles. Yet while useful for understanding the big picture, these expertise studies are often focused on non-scientific disciplinary areas, and those few studies that look at scientists rarely venture into authentic lab or field settings.

Within the geosciences, novice studies far outstrip investigations of experts, and in the few studies of expertise theoretical arguments outnumber empirical research. We do know from philosophical work that the geosciences are much more probabilistic, historical, reliant upon perceptual judgments, and influenced by intuition and imagination than other physical sciences.

Certainly, geological interpretation is by nature historical and uncertain, as geological reasoning reconstructs the most probable sequence of events producing landscape features or rocks. Geologists in the field are physically embedded within their research context, thus geological interpretation can be highly subjective. This is different from more "objective" laboratory sciences (such as chemistry) where the body of the observer is removed from the experiment, but is probably a shared characteristic with other field-based disciplines (such as ecology or anthropology).

Empirical studies of geoscience experts suggest that 2D and 3D visualisation abilities are critical to solving geologic problems, and visual extrapolation of features in rocks and landscapes plays a major role in geological interpretation. This ability to extrapolate visually may be tied to spatial ability as suggested by work in geography. For example, expert topographic map readers move fluidly between maps, the landscapes they represent, and the 3D subsurface structures underlying the landscape, whether actual landscapes or images. Geographers have long investigated the nature of map-reading and mapgeneration expertise using a wide range of data and methods, and recent work has begun considering mapping behaviour in simulated geological settings. Similar investigation of map development in authentic field settings is rare, although at least one study provides empirical evidence for changes in field cognition in response to environmental inputs. In other words, geocognition is situated. Brodaric and Gahegan (2001) also identified the importance of prior experience and knowledge on the development of geological field categories, as suggested by non-empirical discussions.

Methodologies commonly employed by science educators to investigate expert cognition and knowledge structures can be subdivided into four categories of empirical cognitive experiments (recall, perceiving, Categorisation, and verbal reporting). Although these techniques are specifically used in laboratory settings, they can be applied with some modification to authentic settings.

For the investigations of field behaviour, verbal reporting and perceiving tasks are the most useful and easily generated data sets; in addition, behaviour as a reflection of underlying cognition can also be used to generate an understanding of cognition. For example, spatial data of human movement (i.e., navigation) can be used to infer underlying intent and cognitive process. In general, navigation data will be most useful if used in combination with data that can be directly correlated to reasoning.

Verbal reporting and perceiving tasks provide direct observation of subject thinking. Verbal reporting as a specific category of cognitive experiment refers to participant explanations as they engage in a real-time task. These may be Think-Alouds, such as those performed during development of many concept inventories, as explanations of tasks, or during conceptual interviews. Think-Alouds can be used to ascertain both the validity of the underlying task as a cognitive prompt, as well as differences between expert and novice cognition. Perceiving tasks focus on the filtering that occurs as an environment or artifact is perceived, considered, and related to ongoing cognition. These types of data, whether ultimately resulting in verbal reports or artifacts such as notes or drawings, can reveal differences between expert and novice cognitive processes. For example, expert and novice teachers viewing and talking about a video of classroom instruction showed differences in the complexity of observed patterns as revealed in ThinkAlouds. Similarly, research on medical students and professional doctors reveals differences in the amount of material that experts and novices take note of, as well as in the specific material that is considered of interest.

Navigation behaviour can be quantitatively recorded through use of a Global Positioning System (GPS) unit attached to a subject moving through a natural space. Analysis of GPS tracks provides a mechanism for assessment of natural movement; for example, pedestrian way finding behaviour as revealed by GPS suggests interesting relationships between individual and group movement decision-making. Models of natural human movement and geospatial reasoning are computationally intensive when left open-ended, so construction of an activity ontology is often used to identify participant actions throughout the track. For example, an activity ontology could consist of orientation (broad surveying of landscape), movement (rapid or slow progress), and inspection (remaining in one spot). GPS units have been applied to tracking of college undergraduate student movement during geologic mapping by Riggs et al. (2009), suggesting influences of environment on map-making behaviour.

The methods used to engage participants and collect data on field cognition and navigation are described below. We have provided sufficient detail to ensure effective reproduction or modification by other groups.

Research Design: This pilot study was designed to develop and test methods for eliciting cognitive and navigational processes used during geologic field mapping. The original study design included three volunteer participants with varying prior experience in geologic mapping recruited from among colleagues of the researchers; participants mapped at two field locations in Michigan, USA. Data collected from these participants included a background and experience questionnaire, field maps and notes, a GPS track, a Think-Aloud audio log, a photographic log, and a follow-up interview.

A purposeful sample of four students from a college undergraduate field methods course that used the same field locations was later added to the study, in order to obtain more GPS tracks and maps for analysis. However, because of time constraints in the field course, we were not able to collect audio log and interview data from the student subjects. While the non-equal treatment between these groups (volunteers vs. students) certainly affects the outcomes of the project, our original intent was to produce a viable research method and preliminary hypotheses on novice-expert field cognition that could be tested in the future with a larger population of subjects and more controlled research design.

Location: This study utilised two field locations near the town of Marquette in the Michigan Upper Peninsula, USA. Rocks exposed at both sites are distinctive enough that a detailed knowledge of the local geology and stratigraphy was not crucial to effective mapping. At Lighthouse Point (LP) all participants mapped crosscutting relationships between three intermittently exposed rock units onto a simplified topographic base map. The mapping area was 145 metres? 190 metres. At Harvey Quarry (HQ), all participants mapped a cross-section through a structurally complex syncline containing seven rock units. Rocks were intermittently exposed over a horizontal distance of about 450 metres, with the site flagged every 30 metres. Exposure consisted of vertical quarry walls and road cuts; participants drew rock units on a simplified outline of the traverse and were instructed to project rock units and contacts below the ground surface to produce the cross-section. All participants (volunteers and student sample) used the same base map and cross-section outline.

Participants: Seven participants representing a continuum of prior geologic study and field mapping experience ranging from none to 10+ years took part in the study. Participants were initially ranked along the expert-novice continuum based on self-reported prior geologic mapping experience obtained via a written survey. Questions on the GES elicited information about participant demographics (age, gender, race, etc.), education, work experiences, and prior field experiences. Criteria for placement along the continuum included: (1) Number of undergraduate and graduate geology courses taken; (2) Number and level (BA/BS/MS, PhD) of degrees held in geology; (3) Number, length, and purpose of prior geologic field experiences including field courses, fieldbased thesis or dissertation research, field-based work or internship experiences, and teaching experiences; and (4) Years of professional mapping experience, where applicable. Two of the authors reviewed these data and agreed upon the relative level of geologic expertise associated with the backgrounds of each participant. Based on these criteria, we classified participants as five novices (Nl through N5) and two experts (E6 and E7), with participants ranked from most novice to most expert.

Data Collection: The three volunteer participants, including one novice (N2), and two experts (E6 and E7), mapped first at Lighthouse Point and then at Harvey Quarry on two consecutive days. These participants were directed to simply create a bedrock geologic map (LP) or cross-section of the traverse (HQ), to record their thinking during the mapping task with the audio recorder, and to take pictures of interesting features during mapping. Participants were allowed unlimited time, although all volunteer participants completed each task in under 4 hours.

Two continuous GPS track records were used to validate the most appropriate unit setting, with one GPS set to collect one waypoint per minute and the second set to collect data automatically, more frequently when the wearer was moving rapidly and less frequently when the wearer was stationary or moving slowly. Semi-structured follow-up interviews were intended to clarify maps and specific map features, and capture participant reflections of the mapping experience. A set of open-ended probes was used to initiate the interview process. In addition to questions designed to clarify specific map features (e.g., "what does this symbol on your map mean?"), probes included the following questions:

- Is there anything you would like to tell us about today's activity or your map?
- Which parts of your map do you feel most confident about? Least confident about?
- Can you give us an overview of the strategy you used when you did the mapping?
- Imagine you have another day at the field site. Where would you spend your time, and why?
- Imagine you were doing actual research in this area. How might that work differ from what you did today?
- Please take a moment and look at your photos. Can you show us approximately where the most important photos were taken and discuss why you took them?

The remaining subjects (N1, N3, N4, and N5) mapped both field sites one day prior to the volunteers, as part of a two-week, undergraduate field methods course. None of the researchers or volunteer participants were instructors of record for the field course, although one of us (HP) had accompanied and taught portions of the course in past years.

Mapping activities took place about midway through the course, after an introduction to the regional stratigraphy. At both sites, students were explicitly instructed to interpolate lithologie contacts through cover and cultural features, to collect strike and dip of bedding in 5-10 locations spread over the map area, to identify map units by stratigraphie name, and to write a complete description of all map units. The course instructor pointed out several key features and locations to help students begin mapping at the start of the day. Students were allowed an entire day (~8 hours) to complete their map plus additional requirements (e.g., lithologie descriptions); in general, student participants completed each project in 4-6 hours. The two sites used in this study were the students' first graded mapping projects, and were completed independently.

Data Analysis - Participant maps (LP site), cross-sections (HQ site), and GPS tracks were analysed quantitatively using ArcGIS 9.2. Audio logs and interview data were analysed qualitatively using an emergent coding scheme in which themes were not predetermined but instead emerged from analysis of the data. Participant maps were scanned, digitized, and converted to GIS files. Based on our initial examination of the maps, we found several parametres to be the most useful for making comparisons between individual participant maps and for identifying common aspects among maps for experts vs. novices: percentage of area mapped as specific lithological units (rock types), number of mapped units, and average number of mapped polygons per unit.

All GPS tracks were downloaded into GIS and each individual track was examined for total time spent mapping, total distance covered, and average speed. For the Lighthouse Point site, each participant's track was superimposed onto his or her map, and a kernel density map, which assesses spatial patterns of time spent mapping various areas, was created. Relative time spent mapping particularly sensitive areas, such as areas with evidence of crosscutting relationships or contacts, was investigated via this kernel density analysis. A generalised path for each track was created to examine the number of overall loops and retraced paths in an individual's overall mapping strategy, thus establishing individual track complexity. Finally, all tracks for the Lighthouse Point site were compiled into a single kernel density map to establish evidence of trends across the entire participant group. We did not perform this analysis for the Harvey Quarry site as the location required a horizontal traverse and mapping of vertical outcrop to produce a cross-section.

Participant audio logs and interviews were analysed qualitatively using a modified thematic content analytical approach. One author reviewed all participant audio logs and debriefing interviews while examining relevant maps and photographs. Generalisations extracted from participant commentary were complied during the initial review; generalisations included such items as identification of rocks and features, taking measurements, noting location, explicit indicators of mapping strategy or planning, confidence, interpretation of rock or rock feature relationships, obstacles or problems encountered, and making predictions. These generalisations were next grouped into four themes, and subsets of audio files were reviewed by a second author to ensure overarching agreement with generalisations and themes.

We took this condensed approach to inter-rater reliability primarily because only three participants completed audio logs and interviews; this small sample size precludes generalisation of our findings, regardless of steps taken to establish reliability. Finally, while some themes emerged from the data naturally, we recognise that our prior knowledge, field experience, and familiarity with expert-novice literature likely influenced these themes. In particular, as field-trained geologists we expect mapping activities to involve repeated identification of objects and location of oneself on a base map; it is therefore difficult to ascertain if all of the thematic codes emerged naturally from the data or were established a priori.

Quantitative Data: Maps and Tracks - Maps of Lighthouse Point produced by the seven participants were highly variable across the entire population, although six of the maps contained at least some features, such as cross -cutting relationships, in common. Given this disparity, it was nearly impossible to associate a particular lithological unit across participant maps at both study sites. However, it was possible to examine the overall percentage of lithological unit classification in the total mapped area, and in particular locations of interest at the site. It was also possible to examine the complexity of the map by examining the number of units and polygons per unit on each map.

In general, we found that student mappers (N3, N4, and N5) had similar overall percentages of lithological units. We also found total map area devoted to different units was similar for experts (E6 and E7). The two most inexperienced mappers (Nl and N2) produced maps with the least amount of complexity in terms of number of units and polygons per unit. The geology students (N3, N4, and N5) produced the most complex maps, and the two professionals (E6 and E7) produced slightly less complex maps.

Cross-sections produced at the Harvey Quarry site were even more variable across the entire population; in fact, we were unable to clearly identify common lithologie units across all cross-sections, preventing comparison of number of units and polygon locations. We did note that expert E6's cross-section was less structurally complex than all other participants except novice Nl. Expert E7 had the most structurally complex cross-section.

Analysis of individual participant tracks at both sites indicates that participants mapped at very different speeds from one another, and at very different speeds during various portions of the task. For example, novice N2 finished the Lighthouse Point task in just over one hour, whereas experienced mappers E6 and E7 finished in 3.5 and 4 hours, respectively. The remaining students took 4.5 to 5.5 hours to finish the actual mapping plus other course requirements (e.g., lithologie descriptions and final map). Harvey Quarry data show similar trends.

Qualitative Data: Audio Logs and Interviews - Coding of the audio log and interview transcripts produced by subjects N2, E6, and E7 revealed a dozen generalisations that could be grouped into four major themes: navigation and spatial awareness ("Where am I?"), identification of features ("What am I looking at?"), synthesis and model testing ("What does it all mean?"), and metacognition ("How am I doing?").

Navigation and Spatial Awareness: Where am I? - Two key aspects of geologic mapping are the ability to orient oneself in space, either in relation to landscape features or to distance markers, and to negotiate between the actual landscape and the map. Participant comments related to both of these aspects were placed in this category.

All three participants made frequent references to their location in space relative to obvious natural and man made features (coastline, buildings, road, obvious changes in topography, and marker flags). All participants also used orienting techniques throughout each task, particularly at the beginning. Orientation strategies included moving to a high point to get an area overview, walking the entire transect, and changing perspective by moving close to and backing away from the outcrop. For example,

Novice N2: I guess what I am doing now is just trying to take an overview of what the area looks like, get familiar with it since I don't know anything about it. ...just looking at the structures, the topography, and exposures.

Although she made frequent reference to her location in space, the inexperienced mapper (N2) had difficulty negotiating between the map and the landscape, and as a result both of her maps were mis-scaled. She noted these problems in her interview:

Novice N2: 1 had a difficult time finding the flags... It kept throwing me off that the actual outcrop was almost a third smaller than the subsurface area."

Neither of the experienced mappers verbalised this problem. In fact, both experienced mappers allowed aspects of the landscape, such as the scale of the map area and quality of exposure, to dictate their mapping strategy:

Expert E7: I realise that I'm mapping an area that's a few hundred feet long and a couple hundred, 200 or 300 feet wide, and I wonder should I take a walk about to see what I can see? The types of exposures and where the best exposures are. I think the first thing I might do, as far as improving my efficiency would be to walk up at the high point and have a look over the map area.

Although this navigational technique is similar to that used by the novice, this experienced mapper explicitly uses the overall size of the map area to control his strategy.

Procedural and Declarative Knowledge: What am I looking at? -A large portion of the audio log entries referred to identification of rocks and geologic features, and to determining rock and feature orientation. Many common rock types were immediately recognised and named (e.g., basalt, quartzite, slate). When rocks could not be easily identified, participants used additional textural and compositional information, for example:

Expert E6: I'm going to call this a dolomite; it's definitely soft enough to be readily scratched with a rock hammer; it does not react to acid; it's purple. At a glance it looks pretty similar to the quartzite; it's a densely interlocked granular structure of a pale purple crystalline rock but it's too soft to be quartzite and it's got that marly, stromatolitic kind of texture on the weathering surfaces.

Rock features, such as bedding and joint sets, contacts between rock units, and structural features, such as joints and folds, were also commonly noted and mapped. Both expert participants, and to a lesser extent the novice participant, frequently took measurements of rock orientation (strike and dip) as well as estimated thickness of small rock layers. A significant difference between the novice and more experienced mappers was that the experienced mappers often (though not always) recognised rock types and structural features more readily than the novice mapper, who often struggled with rock identification (especially with the identification of weathered rocks):

Novice N2: It was hard for me to identify [rocks] in the field because of the Ward's samples that were given in the lab, which are perfect representations of perfect rocks and minerals. I can honestly say that absolutely none of them... come with algae and weathering and erosion and anything else subject to the environment.

Additionally, E7 included many more descriptions of structural features (faults, shear zones) than E6 in his audio log, which correlates with greater complexity in his map.

Synthesis and Model Testing: What does it all mean? - All participants verbalised interpretation of rocks and rock features to some extent; for example, making mention of rock classification (igneous, sedimentary, or metamorphic), making judgments about subdividing a single unit into multiple ones, speculations about the nature and type of rock deformation, and interpretation of rock origin or deformation history. For examples:

Novice N2: This appears to be an intrusive rock type. Then there's faulting... This area contains a series of fractures and faults. (HQ interview)

Expert E6: 1 still think the granite is intrusive... All of this [granite] is parallel to the two fault planes I measured earlier. So if the granite was synchronous with some kind of regional deformation that might explain the shearing and the intermixing of the rocks here in this transitional zone.

The two expert participants were far more likely to verbalise interpretation of map features in the audio logs, and both experienced mappers tested their tentative mental models against additional observations:

Expert E6: ... I thought, well, it was either a syncline or some kind of repeated section because of faulting... I'm not sure that these strata [south side] duplicate these strata [north side]. (HQ interview)

Expert E7: ...I looked around, but the dominant lithology... was... quartzite, so I'm wondering what happened to the slate? ...so I speculated that it's a fault. (HQ interview)

Metacognition: How am I doing? - Metacognition is generally defined as knowledge of one's own cognitive processes and includes self-awareness and regulation of the learning experience. All three participants made metacognitive comments in both the audio logs and interviews, mainly as they described their mapping strategies. Both the novice mapper and more expert E6 describe a strategy of looking over an area, determining the rock types and structures, and then look at the same area from a different viewpoint to evaluate the initial opinion:

Novice N2: I just did my best to scan the area and try to find any samples or anything that I could do to help me describe what the rock types were and what was going on. So I would scan broadly and then narrow it down and then if I moved it to another area I would scan again and go narrow. And then... when I was working my way back, I would still look back over the areas that I had already mapped just to make sure my point of view wasn't going to change my opinion of what I had done. And I did that with each spot that I went to. (LP interview)

Expert E6: My general game plan here is to just start at one end and approach this pretty linearly, just keep walking to the north, take strike and dips, take lithologie descriptions, and see what I see as I work my way just along the face linearly. (HQ audio log)

Metacognition also includes one's awareness of distractions to the task at hand. All three subjects pointed out distractions to the task, although overall the novice mapper made far more frequent comments about physical and psychological distractions than the experts. For examples:

Expert E6: All righty... I lost my acid bottle somewhere. I made it to about [the] 1,000 foot mark there and realised I might have needed it and I've got to backtrack now and try to find it. (HQ audio log)

Expert E7: I walked across the highway, which is the most dangerous thing I've done today. (HQ interview)

Novice N2: The area that was covered with shrubs... was kinda creepy and I stepped on a dead animal... so I didn't like going through that area. And then this area... was incredibly close to the highway and pretty steep, so I didn't feel incredibly comfortable at that end. (HQ interview)

The mixed methods approach presented here provides useful insights into geocognition when data are triangulated; multiple analyses on some data, such as the maps, are much more insightful than a single analysis would have been. The collection and analysis of GPS tracks offers evidence of actual participant movement during mapping, while Think- Alouds offer tangible links to cognitive processes underlying behaviour. We provide below some preliminary suggestions for collection of each of the data sets utilised here. In addition, we discuss findings about expert-novice strategies as suggested by these data, with the caveats that the participant population was small and non-random, treatments were non-equivalent, and thematic codes may or may not have authentically emerged from the data. We follow this brief analysis with recommendations for application of this methodology to more robust geocognitive investigations of experts and novices in authentic field settings.

Findings: Methods - This study enabled us to test methods of collecting and analysing a variety of data sets. Based on our experience and preliminary findings emerging from the data, we can make some recommendations for future work of this type. Due to the wide variety of maps and cross-sections produced by the participants, we found that percentage of area mapped as specific lithological units (rock types), total number of mapped units, and average number of mapped polygons per unit provided the most useful analyses. In order to make more meaningful comparisons between participant maps, we recommend that all participants share the same familiarity with the study area and the same expectations for naming of map units. We also propose that giving participants a standard way to indicate certainty on their maps (e.g., solid lines for certain contacts, dashed lines for interpolated contacts) would also enable more meaningful comparisons. Although purposeful samples, such as that used here, are a common mechanism for easily obtaining novice participants, we recommend that all subjects in novice-expert studies be recruited in the same manner.

The ArcGIS analyses of the GPS data provide invaluable insight into participant movement. We found that the automated GPS setting was able to capture finescale movement of the participants through the site; however, the non-equal time increments brought unexpected difficulty in determining appropriate statistics for temporal analyses. We did find there were clear advantages to the automated setting, including a significant reduction in the size of the data set. The automated setting also defined participants' stops in a uniform and automated way, rendering this aspect of the analysis much easier. Overall the automated setting was preferred over the 3-minute GPS record used by Riggs et al., (2009). By itself, the GPS data provide evidence of subject movement; this movement can be fully understood when linked to audio data suggestive of underlying cognitive processes.

Analyses of Think-Aloud audio logs and follow-up interviews suggest that these are a powerful means of elucidating cognitive processes in the field for both novices and experts, and provide the cognitive link necessary for fully interpreting navigation data. Our volunteer participants were instructed to simply record their thought processes as they were mapping; all three participants recorded ~20-50 total minutes of comments during mapping (total mapping time ranging from just over 1 hour to just under 4 hours), suggesting that many thought processes were captured. For future work, we recommend that audio comments be time stamped, so that portions of the audio log can be exactly and easily correlated with GPS tracks.

In addition to the Think-Aloud audio logs, follow-up interviews proved important for clarifying maps and strategies. For example, we were able to ask participants to clarify their map symbols. We were also able to explicitly ask about participant certainty related to the map produced and about particular strategies used to create the map.

We recommend that in future work, all subjects participate in a follow-up interview immediately following the mapping experience.

In general, we noted good agreement with participant recollections of the map experience as expressed in the interviews and in the actual real-time audio logs. Overall, the audio logs captured many fine-scale thought processes that were not evident in the interviews. For example, expert E6 mentions in the interview that he worked out the cross-cutting rock relationships at the Lighthouse site by going to two critical locations.

Yet the audio log reveals exactly how he worked out the age relationships of the rocks, the clues he used, and that he revised his opinion several times before drawing a final conclusion. These fine -scale thought processes would not become apparent from map, track, and interview data alone. Given the effort required to collect, transcribe and analyse audio log data (estimated at 10-20 hours per audio log), we recommend that the specific research questions of importance to the researcher should dictate whether interviews alone (which capture generalised thinking) or interviews coupled with audio logs (which capture moment-by-moment thinking) would be most appropriate and necessary for probing thinking in the field.

Findings: GPS Tracks and Maps — At the Lighthouse Point site, the analysis of map complexity (number of units and polygons per unit) revealed that complexity was the greatest for those participants we identified as intermediate along the expert-novice continuum (i.e., the geology students had the most complex maps). We also found a high degree of similarity between the proportion of units mapped as particular lithologies.

The field course students (Nl, N3, N4, and N5) were taken to several sites in the study area and shown rock types and cross-cutting relationships by their instructor. We therefore speculate that these differences may be due to the information received by students from the course instructor, rather than necessarily an expertnovice difference. However, comparison between the remaining participant maps (N2, E6 and E7) reveals some interesting expert-novice differences. At Harvey Quarry, expert E6's cross-section was less structurally complex than expert E7's cross-section. We speculate that E7's prior mapping experiences may have made him more aware of structural features, as well as more likely to interpret rock features as structures than E6.

Our analysis of the GPS data suggest that collecting waypoints under the automatic setting captures fine-scale movements of the wearers, enabling a more accurate log of participant locations to be captured and facilitating some analyses. This is probably similar to collection of time-differentiated data at much shorter time intervals (tens of seconds) than the three minutes intervals suggested by Riggs et al. (2009); this suggestion will be tested in future work.

Kernel density analysis of GPS tracks indicates that certain portions of the landscape were likely to be covered multiple times by a single participant, while relatively few participants attempted to examine those areas that were difficult to access. Analysis revealed several broad categories of participant movement; for example, some participants spent a relatively long time on the catwalk crossing the centre of Lighthouse Point, while other participants spent the majority of their time walking the outside edges of the landscape. Some participants mapped quickly "on the fly" and then sat for long periods to draw maps and compile notes, whereas others drew and described as they walked through the area. Additionally, some participants utilised a particular strategy (example: walk perimetre completely twice, revisit locations of complex relationships), while other participants walked much further distances, backtracking several times in large loops and seemingly at random. Overall, kernel density analysis suggests that experts (E6 and E7) display economy of movement during mapping, as suggested by simpler movement tracks and less backtracking than novices. These two experienced mappers also spent more time in key areas, such as near contacts, than novices. These map and track data suggest that underlying differences in participant experience, background, and mapping strategies will influence field behaviour and interpretations.

Findings: Audio Logs and Interviews — The ability to produce a geologic map requires interpretation of the rocks and structures that are present in the map area; in other words, to synthesize the rock and rock feature data into a model that consistently explains observations. Interestingly, we found no evidence that the novice participant synthesized observations into a model, correct or incorrect, of Harvey Quarry (a syncline) on either the Audio Log or during interview probing designed to encourage synthesis. For example:

Did you see anything that might be called a structure? Novice N2: This appears to be an intrusive rock type... Then there's faulting...

This area contains a series of fractures and faults... it seems to me that these was some serious uplift... but I don't know if I can generally say there's one thing going on. (HQ interview)

Conversely, the two expert participants were far more likely to verbalise interpretation of map features in the audio logs; additionally, both mapped the quarry site as a syncline and were able to explain how they reached this interpretation. In addition to synthesizing data and forming mental models that explain the observations, both experienced mappers tested their tentative mental models against additional observations. Interestingly, each expert had a different model-testing strategy. Expert E6 spent a significant amount of time understanding contact relationships and how rock units fit together; he made initial observations followed by later interpretations, and although he found additional data that did not fit his model, it was insufficient for him to immediately change his model:

Expert E6:...it was only later, after I made it all the way to the north end and started working my way back that I started to realise all the strikes and dips were pointing toward the centre, and I thought, well, it was either a syncline or some kind of repeated section because of faulting. So basically descriptive in the beginning and interpretive toward the end. In general I drew a syncline. Not 100% positive that that's accurate, because there's a lot of cover and this is all weathered really recessively in here, and I'm not sure that these strata [south side] duplicate these strata [north side]. So this might be better explained as a structure, like a fault instead of a fold. (HQ interview)

Expert E7 utilised a different model-based strategy, verbalising interpretations of the origin of rocks and structural features during mapping, as well as attempts to infer the geologic history of the site. He also fit together observations into models, but in contrast to E6 he made initial models to test and often changed his model based on new evidence:

Expert E7: Continuing back into the woods, there was a beautiful outcrop of stromatolites that are essentially horizontal... so I was speculating that I was in the middle part of a syncline... As I came out of that... I was confused because if this was truly a syncline, then I should come out of the basal part of the stromatolite and back into the slate. I looked around, but the dominant lithology... was... quartzite, so I'm wondering what happened to the slate? ...so I speculated that it's a fault. (HQ interview)

When E6 expressed confusion about the rock types, location of contacts, and his interpretation of the rock structure (the syncline), he revised his strategy:

Expert E6: I'm going to change my strategy a little bit here. I'm going to walk all the way to the northern end and start working my way back. The middle zone here, where it's weathered recessively and heavily vegetated, I'm going to come back to; I'm going to save for last. So I'm going to, from here just walk to the far north end and then start backtracking my way towards the centre. (HQ audio log)

All three participants made frequent comments expressing degrees of certainty about their location in the map site, rock and rock structure identifications, and interpretations of the rocks and structures (their mental models). Comments ranged from being "sure" of a particular location, rock type or feature (e.g., measuring a bedding plane), to whether an observation "makes sense" in light of other observations, to "just guessing" about a rock type or conclusion.

All of the participants could point out areas of their maps that they felt certain and uncertain about, and could articulate reasons for their degree of certainty. However, each participant was unsure for different reasons. The inexperienced mapper (N2) expressed the most uncertainty over rock type identification. Participants E6 and E7 both felt confident of the rock types they mapped, but were unsure of contacts and structural features; for example, participant E6 wondered about the overall syncline structure he had mapped at the HQ site, and participant E7 felt unsure of his identification of several faults at that same site.

Overall, key differences between novice and expert mapping were apparent, and these resulted in both bedrock maps and GPS tracks that varied in complexity. Based on interpretations of the audio log data, the novice mapper was far more likely than the two experts to struggle with determining her location, negotiating between the map and the landscape, and identifying rocks and rock features. She also noted many more distractions to the mapping task. More critically, the novice mapper did not synthesize data to form a mental model of the rock relationships at the mapping site, nor did she test her model against additional observations. We recognise that one subject cannot adequately "represent" all novice mappers. However, anecdotal evidence, based on personal experience teaching a field methods course and conversations with colleagues who teach field mapping, suggests that the difficulties described by the novice mapper (rock identification, location, synthesis of data) are typical of students new to mapping.

Although the two experts had different approaches to developing and testing models, both could clearly articulate the strengths and weaknesses of their models and, consequently, their final map. Additionally, both experts made frequent reference to the certainty of their observations and interpretations, and both adjusted their mapping strategies to account for new observations or their own new insights. We conclude, therefore, that metacognitive aspects (e.g., planning strategy, monitoring performance, revising strategy, staying on task, and evaluating the final product) play a perhaps previously unrecognised but vital role in field mapping performance.

We speculate that the experts' ability to monitor and evaluate their performance contributed to how they planned their movement through the map area to take advantage of better rock exposures and recognise key features. In addition, the GPS tracks and complexity analysis, coupled with audio logs, suggests that mapping strategy may impact map complexity.

Given the mixed volunteer and purposeful sample used in this study, we chose not to evaluate participant maps for correctness relative to a published map or expert-provided answer key. In the future, comparison of maps generated under equal treatment conditions to a published map would be advantageous. In addition, comparison of map correctness and Think- Aloud data, as cognitive proxies, to GPS tracks, as navigational proxies, will generate a deeper understanding of the influence of cognition on behaviour, and vice versa, in field settings. Overall, these findings agree with the limited existing work, such as that of Brodaric and Gahegan (2001), suggesting that geocognition in field settings is situated, and that cognitive processes are tied to both environmental inputs and prior knowledge.

Results of this study suggest that scientists are quite varied in how they perceive and process data, and that geologic data are perhaps more varied by perception of the individual scientists than we like to believe. For example, at the quarry site, expert E6 was more inclined to treat lithological boundaries as contact relationships whereas expert E7 was more inclined to treat them as structural discontinuities. Although both experts mapped a syncline, the two maps differed significantly in structural complexity as a result of these personal biases. Although we would like to think of geological data, and in particular geological maps, as objective representations of the surface and subsurface, they are perhaps more laden with personal interpretation than we care to admit.

Our results also suggest that in addition to teaching the mechanics of field mapping (i.e., note location, identify rocks and features, take measurements), field mapping instruction should focus on the metacognitive aspects of learning as well. Explicitly teaching metacognitive aspects of problem-solving in physics, for example, has been shown to improve students' problem-solving skills as well as their conceptual knowledge of physics.

In particular, geology students may benefit from explicit instruction in how to take observations and synthesize these into models, and how to test the models against additional observations. Explicit instruction in planning and self-monitoring could be beneficial as well; for example, one might teach students how to use aspects of topography in planning a mapping route. Actual, physical modelling of the metacognitive aspects of mapping practice may help students understand these key components of successful mapping.

Future work will replicate this study on a larger scale with participants representing many levels of expertise, including undergraduate students on their first mapping experience, advanced students who have completed field mapping courses, and professional geoscientists with a range of mapping experience.

We will look to support tentative conclusions described above; while this study provided a rich source of information on geologic mapping strategies and behaviours, we do not yet have enough data to distinguish between patterns indicative of expertise in mapping and patterns idiosyncratic to individual participants.

We also plan to investigate how mapping performance varies with expertise in different areas of geology, in order to further determine how expertise "colours" how geoscientists perceive and interpret rocks. Finally, we expect that this line of research should lead to a better understanding of cognitive processes needed for effective field mapping, and ultimately will point us towards mechanisms for engendering these processes in students.

Using Student-managed Projects

Kerka (2001) defines capstone courses as "culminating experiences in which students synthesize subject-matter knowledge they have acquired, integrate cross-disciplinary knowledge, and connect theory and application in preparation for entry into a career." Capstone courses better prepare students for the professional world where strong interpersonal and team skills are highly valued. One of the main characteristics of the current generation of students is that they are team-oriented. They are more interested in working with each other on a project than individually.

Capstone courses in the geosciences typically revolve around individual senior research theses or summer field camp. There are few examples of team-oriented capstone courses in the geosciences. In contrast, team capstone courses have a long history of use in the different fields of engineering education. These courses are typically the engineering senior-design courses, where a small group of students perform hands-on work on real or made-up projects for real or fictitious clients. However, few of these courses involve active management of the project and team assessment by the students.

Having students manage the project creates a sense of ownership of the course in their minds. That ownership affects student attitudes because it gives students an enhanced perception of influence and control: they become more personally involved with projects, they learn to work more effectively in groups, ana the students achieve learning objectives. Team assessment also increases the sense of ownership by further attempting to simulate an actual working environment, where peer supervisors evaluate the performance of individuals within the team, a practice that occurs less frequently in science courses that in other subject areas.

The theme of the capstone course is Environmental Impact Assessment (EIA). The United Nations Environment Programme (1987) describes EIA as an examination, analysis, and assessment of planned activities with a view to ensure environmentally sound and sustainable development. EIA is a methodology to examine the environmental costs and benefits of development projects when they are proposed, rather than after they are constructed. In the US, federal and private organisations carry out Environmental Impact Assessments to support federal or federally funded actions. In addition to federal agencies, over 30 states have legislation requiring similar procedures. Many regional planning organisations and local municipalities have EIA requirements for land-use planning and zoning decisions. The result of an EIA is a report called an Environmental Impact Statement (EIS) that details the best course of action or the best alternative with respect to the environment to execute a proposed action.

My efforts to use a student-managed, project-oriented capstone course as a teaching tool have been adapted and modified from the experiences of others, especially from engineering senior-design courses. Previously, this course consisted of small made-up projects that teams of three or four students would work on to produce a report that was graded by the instructor.

The course did not engage the students, did not involve the community, and did not give students the tools to enter a career. It covered the academic material, but was not a true capstone course. The previous course lacked the engagement and challenging components of a capstone course. The projects were not relevant because they were only made-up academic exercises that included no one outside of the small team of students preparing each report. The students were not challenged to extend their university-learned skills as the report constituted just one more of the typical term-projects from other courses.

The new course described here is both engaging and challenging. It requires the whole class to work as one team on a real project requested by a client outside the university. The client is interested in the results of the student-managed project, the EIS, and many times this final report decides what the client's course of action will be. The students are challenged by having a client they have to respond to and by having their work influence a decision in the community. They are also challenged by having to work together in a team managed and assessed by their peers and not the instructor.

Class Structure

The capstone class meets twice a week for 75 minutes each time. The length of the course during the semester is divided roughly into equal thirds. The first third briefly covers the history, philosophy, and legal authority for environmental impact assessment. This part of the course is based on a combination of lectures and discussions from assigned readings from several sources and the textbook. Specific assignments for this section of the course include summaries and reviews from actual Final Environmental Impact Statements (FEIS) produced by federal agencies and easily found in the documents section of the library or the federal agencies' websites.

During this part of the course I use varied active learning techniques that stress team work and cooperation, such as think-pair-share, jigsaw classroom and gallery walk. These active learning techniques serve two purposes for students in this class: to have more control of the course and to start working with each other.

The second third of the course focuses on methodologies and techniques for Environmental Impact Assessment (EIA). This section is strongly tied to the capstone project. Each methodology and technique is presented with respect to the specific capstone project of the course, and each homework assignment directly applies to the progress of the capstone project.

At the beginning of this section, I reserve two lectures to bring all the team members up to speed with respect to one or more technical aspects of the capstone project. For example, if the project is related to parking lot construction, then we discuss rainfall-runoff processes and the impacts of impermeable surfaces.

The last third of the course is devoted exclusively to the capstone project. No new material is introduced in this part of the course. The class periods are used only for team meetings to discuss progress, assign new work, and keep everyone on the team on track. During the last week of the semester the students present the results of the EIA project orally to the client and to interested community members, and in written format (electronic copy of the Environmental Impact Statement).

The student-managed project is the core of this capstone course. Except for the instructor's task of finding the project and framing it with respect to an Environmental Impact Assessment, the students carry out, organise, supervise, and assess all project work. The course instructor only serves as a consultant to the students. The whole class works on only one project during the semester. The deliverables to the client are a written copy of the Environmental Impact Statement, detailing the best course of action for the project, and a copy of the electronic file used for the oral presentation.

The project for this course varies every semester with respect to the location (environmental setting) and the type of project. In contrast, typical project-oriented or capstone courses in the geosciences utilise the same physical location (e.g. a watershed) on a continuous basis.

Projects for this course have included environmental impacts of siting wind turbines, constructing porous parking lots, switching a vehicle fleet to alternative fuels, and constructing an interstate truck plaza. Although these projects would not generally require the development of an Environmental Impact Statement (EIS) due to their reduced magnitude and significance of impacts, they are good examples of the kinds of issues occurring in project management. Once our students graduate, they will likely face similar situations at work.

The usual enrollment in this course is around twelve students, though it has been as high as eighteen or as low as ten students. The project team is assembled by electing a project manager and two project coordinators.

The rest or the students become part of any of two teams under one coordinator. In my experience, this tall structure versus a flat structure (only a project manager) works better at distributing the work and delegating responsibility among all team members. The better functionality of a tall structure may also be due to limitations of an inexperienced manager in handling a large team. When using only a student project manager and no project coordinators, the project manager got overwhelmed very quickly as the tasks quickly accumulated. Having two project coordinators and no project manager created some decision-making conflicts and the look of the final report was not uniform.

The project manager and two project coordinators are elected by voting from all the students in the class. It is typically done with a secret ballot. Any of the students in the class can be elected. However, any student can be excused from becoming a project manager if they think that they will not be able to perform the duties due to other time commitments such as a heavy academic load or demands of current job. Project coordination demands less time than project management and typically students do not need to be excused from becoming one. Obviously, more time spent on the class has to be rewarded and the project manager gets a 5% bonus on the final grade for performing this duty. The election takes place approximately one month into the semester, once the students have the chance of meeting and working with each other. This is a capstone class, so most of the students have taken classes together before and already know one another.

The duties of the project manager are to be the point of contact with the client; to keep me client informed of the progress and to make sure that all the client's issues are addressed, and to put together the final document (EIS); making sure it has a unified look and feel, since the writing is done by all the team members. The project manager also works with the project coordinators to find the best way to subdivide the project into tasks and distribute them to the team members. Some projects work better by grouping the team members by knowledge and skill and others by grouping them by work process and function.

The project coordinators' tasks are to distribute the work to the team members, deal with the every day issues of managing the teams, and compile the work done by the team for submission to the project manager. The project coordinators, together with the project manager, create the oral presentation from the information gathered by all the team members, and the coordinators present it to the client and community.

The instructor serves as a consultant to the project manager and coordinators as well as to any team member. This consulting role may involve how to deal with people management issues or detailed technical questions. My role as a consultant is not a passive one. I stay in contact with the project manager and coordinators to make sure that the project progresses smoothly. I attend all group project meetings, but only as an observer and usually take a seat in the back of the room. In contrast to the availability of the consultant (instructor) at all levels, the project manager is the only student with direct contact with the client.

Working with an actual client and project makes this course a capstone experience, not just an academic exercise. The clients have a genuine need for the results of the project. The report prepared by the class will help in the decision making process of the client with respect to the project (an environmental impact assessment determines what is the best way to carry out a project). This aspect of engaging in a "real world" project as well as the realisation that the team work will have a direct impact on how a project may proceed really excites the students. They work hard not only for a grade, but because their work can make a difference outside of the classroom.

Peer Assessment

The content of an Environmental Impact Statement (EIS), the written report, is well defined. For purposes of assessment and report integration, the main project load is typically divided into four or five homework assignments that correspond to the different sections of an EIS. As each section is completed by a team or teams, it is turned in to the project manager and instructor for review. The project manager and instructor review each section and give feedback to the appropriate project coordinator if information is not thorough or complete. The project coordinator will then assign the revision to a team member or may revise it himself/herself.

However, the project coordinator may not have any enforcement authority if part of the assessment does not reside in that position. Together with each homework assignment (completed section) the project coordinator submits a rubric to the project manager and instructor assessing the work of the team members for that section. The rubric is handed out before the project starts, so the team members can better understand what is expected from them. This periodic assessment helps the team improve their skills as they go. Peer assessment encourages the students to become more involved in their own learning process and repetitive assessment is a necessary part of it.

The grading by the coordinators constitutes a large part of the final grade of the members of the team in the project, and since the project makes a large part of the course, a large part of the final course grade. The team members also have an opportunity to assess the work of their own coordinator.

However, the coordinators' assessment only takes place at the end of the course. I have debated not doing a repetitive assessment for the coordinators. My concern is that having mutual and continual assessment of both parties (team members and coordinators) may influence the objectivity of the evaluation. Nevertheless, throughout the project I am in contact with the project coordinators and am able to give them feedback in their performance. The project manager is assessed by the project coordinators at the end of the course.

Transferability

We recognise that not every capstone course can be turned into a student-managed project due to the size of the project or the nature of the material covered in the class. However, many capstone courses are project-based and turning over me management and assessment of the project to the students can be a very rewarding experience for the students and the instructor. Finding a "real project" has not been difficult. However, it requires the instructor to spend time before the semester starts talking with possible clients to find a suitable project. It is always advisable to have a back up project in case the main project does not go through, is not approved, or gets cancelled.

Clients for the course have included municipal utilities, regional planning commissions, city parks departments, and even our own university administration. It is my experience that once the surrounding community, especially the city and county authorities, and NGO's (non-governmental organisations), realise that there is a group of students at the senior level eager to work on some aspects of project development (EIA) the list of possible projects increases. So, the instructor of this course also has a public relations function.

It is my intention to gradually evolve this class into an environmental management tools course, where EIA would be just one of the tools. This would give more flexibility with respect to the type of projects that could be addressed each semester in the course such as environmental site assessment, life cycle assessment, environmental auditing, ecosystem-based management, environmental risk management, etc. I am also interested in integrating into the capstone course a junior-level course currently being developed in field methods in environmental science.

This new field methods course could serve as a "subcontractor" for some of the field work measurements needed for the capstone project. Currently, most of the research done for the EIA is bibliographical in nature due to the double constraints of time and expertise. However, with the use of some of the time of the field methods class, research in the capstone course could be more dependent on actual data collected at the project site.

The pedagogical benefits of a student-managed capstone course are many. Students learn to think through and resolve problems on their own, which is a skill that many science students are missing. The experiential learning of this course, where students learn by doing, actively involves and empowers them through a "bottom-up" versus "top-down educational process.

The classroom becomes an environment where students make the decisions through applying what they have learned. This course also better prepares them for the professional world, where strong interpersonal and team abilities are highly valued, by making group work among students more effective. For the project manager and coordinators, it provides them with an added set of managerial skills. These are not learned in the typical classroom setting and are eagerly sought in the professional world.

Engagement is another pedagogical benefit of this course. The students provide a service (the EIS report for a proposed project) to the community (the client from the municipal utility, regional planning commission, or city parks department) and in return receive experiential learning. The students nave the opportunity to contribute to their community in a concrete way and learn valuable professional skills. The majority of the students' comments on the class reflect their enjoyment of a course that applies the concepts rather than just reading about them, and their high regard for the practical experience gained, as well as having a say on a project that affects the community where they live.

The students' experience in a capstone course can be greatly enhanced by transferring the management of the capstone project to the students. My role as an instructor then becomes more of a designer and facilitator of learning experiences and opportunities rather than an imparter of knowledge. Returning the control of the course to the students, especially when their work is going to be reviewed by someone outside of the classroom (the client), can be somewhat frightening. However, in my experience with the different groups of students that have participated in this course, the students always perform at or above expectations because they are interested in the outcome of the project and want the approval of the client. This is reflected by the comments and decisions with respect to the projects that the clients have made (i.e., following the recommendations of the project report, assigning other projects for this course to work on, etc.)

This course has benefited the students by increasing the number of skills they have to enter a career. It better prepares them for the workforce. The course has affected me, the instructor, by giving me a new appreciation for the effort and responsibility of the students when challenged with a concrete project that influences the world outside the classroom. So, it begs the question, are we challenging our students enough with our courses?

Numerical ground-water modelling is a critical component of modern hydrogeology but is typically given only a very superficial treatment in undergraduate hydrogeology courses. One reason instructors must often gloss over ground-water modelling is that the software can be extremely frustrating to learn. By the time students learn how to run the ground-water modelling software, little time remains to learn the science.

Alternatively, theoretical discussions of modelling can leave even the most computer-oriented student disinterested without a hands-on component. In the web-based curriculum described here, we try to retain the hands-on aspect of ground-water modelling, while discarding the baggage of learning a real ground-water modelling program. The student is guided through the conceptual exercise of building a ground-water model, and then calibrates the model to actual data in a virtual three-dimensional environment.

The objective of modelling ground-water at a given locale may be to test specific hypotheses, predict future behaviour, or simply to organise available hydrogeologic data in a consistent conceptual framework. The modelling process is somewhat independent of the objective. The modeller must define a purpose, develop a conceptual model based upon available field measurements, select the appropriate modelling software for the stated purpose, design the numerical model, calibrate the model to measured data, and then present the results. If the calibration is not acceptable, i.e. the simulated data do not match the measured data, then the numerical or even the conceptual model might have to be revisited. This processes is repeated until the model is deemed satisfactory.

Note that the only step that requires detailed knowledge of the numerical methods is the design of the numerical model. In a finite difference or finite element model, this means selecting a computational mesh, defining boundary conditions, assigning the correct mathematical equations and constants to each cell in the mesh, and determining appropriate time steps. If the conceptual model is properly constructed, the model design stage should be primarily an exercise in overlaying the conceptual model on a computational grid.

Translation of a conceptual model to a numerical model requires some knowledge of the numerical methods to be used and the particulars of the modelling program. Each program has its own peculiarities that require familiarity to overcome. For example, in the industry-standard MODFLOW model, each computational cell must be specified as hydraulically confined, unconfined, or be forced to depend upon the head solution at each solution iteration. It would seem that the simplest option is to let the program decide whether the cell is confined or unconfined, but in practice choosing this option can slow convergence to a solution or cause instabilities that may prevent the program from converging to a solution at all. These numerical subtleties are important in real applications, but they have more to do with numerical methods than ground-water.

We attempt to separate ground-water modelling as a concept from ground-water modelling as a numerical exercise. Our approach is to guide students through the ground-water modelling process ignoring numerical methods as much as possible. Students are asked to make a series of critical decisions that lead to a numerical model and then to calibrate the model to actual data. To avoid having to wait for numerical calculations, the model results are pre-generated so that students receive instant feedback from their trials. Model results are checked against scientific expectations and field measurements.

Measured data and model results are displayed in a virtual three-dimensional (3-D) environment. This is accomplished through Virtual Reality Modelling Language (VRML) datasets that allow students to manipulate the terrain as a 3-D object, or even fly or walk through the topography. Presentation of model results in this manner enables students to more readily compare simulated results with their own intuition. In addition, hydraulic potential surfaces can be displayed without the need for maps. Hydraulic potential contours can be frustrating for introductory-level students, and these frustrations distract from understanding the underlying principles of ground-water flow.

This lesson is part of an on-going Internet-based educational project called the Mirror Lake Interactive Teaching Database (ITD).

The overarching purpose of the project is to distribute over 50 years of hydrologic data collected at the Hubbard Brook Experimental Forest. Our efforts are directed at distributing these data in a more organic format than individual publications and tables. Data are offered in VRML format, presented in a historical and hydrological context, and integrated with educational lessons add value to the information.

The Mirror Lake Site and Ground-Water Model

This lesson concerns the Mirror Lake Watershed, located within the Hubbard Brook Experimental Forest, in the White Mountain National Forest, near North Woodstock, Grafton County, New Hampshire. The Hubbard Brook Experimental Forest is a 3,160 hectare reserve established by the USDA Forest Service in 1955 as a major centre tor hydrologic research in New England, and is a National Science Foundation Long Term Ecological Research Site. Over 40 years of atmospheric and surface water data have been collected continuously at the site. Since the 1970's the United States Geological Survey (USGS) has been conducting ground-water studies in the Mirror Lake area. In 1979, efforts to understand the interactions of lakes and ground water led to long-term gaging of streams, piezometres, bedrock wells, and weather stations in and around Mirror Lake. In more recent years, additional research by the USGS involved fractured rock hydrogeology.

The study area occupies about 10 km^2 composed of glacial mountain terrain with steep hillsides and flat valleys. Most of the land surface is covered by glacial deposits consisting primarily of till with some local areas of sand and gravel. Glacial deposits are thin or non-existent in the uplands, may be as thick as 50 m near Mirror Lake. The highest point, located in the Northwest corner of the area, is about 700 m above sea level, while the lowest point is along the Eastern edge at about 180 m. Water in the study area drains to Hubbard Brook or Mirror Lake. Mirror Lake drains into Hubbard Brook and, ultimately, the Pemigewasset River.

The surface-water basin of Mirror Lake has changed in area since the construction of 1-93 and is divided into three sub-basins. The total basin area was 1,030,000 m^sup 2^ prior to the construction of 1-93. After completion of the interstate, an earthen dam was built to divert road-salt contaminated surface water away from Mirror Lake. This dam, along the west side of the highway, cut the basin area to 850,000

m^2, diverting water North. The ground-water model discussed in this lesson is based upon a MODFLOW model published by the United States Geological Survey (USGS). Input files and MODFLOW code are available through the USGS, and were supplied to us by the authors. The model grid extends somewhat beyond the Mirror Lake surface water basin. The vertical domain is divided into five model layers. Layers 1 and 2 coincide with the glacial deposits, and layers 3, 4 and 5 coincide with the bedrock. Modelling of this system has shown that the ground-water drainage basin extends well beyond the surface-water basin at Mirror Lake.

The Ground-Water Modelling Lesson

Students are expected to enter with an understanding of Darcy's Law, concepts of hydraulic head, and some general understanding of ground-water flow. The lesson is, therefore, appropriate for the latter part of most introductory hydrology or hydrogeology college-level courses, and perhaps some K-12 Earth Science courses.

No specific computer or mathematical skills are required for the lesson, other than facility with a PC and point-and-click drawing. The lessons begins with a short introduction to the fundamentals of ground-water mass balance. We review the concepts of topography-driven ground-water flow. Water enters the subsurface through recharge, and generally exits through ground-water discharge at surface-water bodies. The height of the water table depends upon the resistance to flow, which is called hydraulic conductivity. With this simple explanation, students are well on their way to understanding ground-water flow modelling from a practitioner's point of view.

The purpose of the model is to determine the average hydraulic conductivity of the glacial deposits and the average ground-water recharge within the study area. There have been slug-test estimates of hydraulic conductivity of Mirror Lake glacial deposits, but point measurements do not necessarily represent regional behaviour. A model-derived estimate of hydraulic conductivity is probably more representative of the watershed as a whole.

In the development of the conceptual model, we stress the importance of proper boundary specification. Two types of boundaries are considered: (1) a perimetre boundary that constrains the entire model, and (2) internal boundaries that constrain localised areas within the model. It is the former that is most important, because it determines the scope of the study and greatly influences its outcome. The students investigate how hydraulic no-flow boundaries may be established in a ground-water model by identifying points where flow lines are expected to converge and diverge. At convergence and divergence points ground-water flow is primarily upward or downward and thus no water flows laterally in or out of the model domain.

Students can investigate how hydraulic boundaries are established by working an exercise using a Java-based finite element program developed by the U.S. Geological Survey called TopoDrive, that calculates flow lines through topography in profile view. They are asked to develop a counter-example, in which hydraulic divides observed in shallow piezometres are found to have only a local influence. The purpose of this exercise is to emphasize the point that hydraulic boundaries are tenuous and established primarily out of convenience.

Once the students understand how no-flow hydraulic boundaries work, they are asked to find them in their model domain. The students are asked to draw a proposed perimetre boundary for their conceptual model based upon a topographic map of the area. Although we include a software that allows them to carry out this exercise on the computer, in classroom testing we found that the students were much more comfortable drawing on a paper-printed map.

The software is a very straight-forward image program, but it takes some time to get used the mouse behaviour and it is difficult to see the whole model area on the computer screen. We now offer the option of printing the map or working online.

Model design is the process of translating the conceptual model into a numerical format by specifying grids, establishing boundaries, and providing model parametres for every computational cell. We try to tread lightly through the discussion of numerical design because the treatment of these details is largely program-specific. Grid design, for example, differs greatly between finite-difference and finite element programs. Our goal is to discuss modelling relatively independent of software.

Although we avoid attention to the particular grid method used, we devote some discussion to mesh refinement as a concept. Mesh refinement refers to the increased resolution of grid near areas of interest. For example, almost all of the available well data are clustered around Mirror Lake. The students are encouraged to think about where spatial resolution of measured data warrants mesh refinement how this refinement affects the level of detail in the model output. We emphasize this non-uniform distribution of measured data by offering a VRML rendering of the wells completed, in the watershed. The wells, topography, and hydrography are shown in 3-D, so that students can easily discern the density of data with depth and with position.

Calibration of the model is carried out entirely in a VRML environment. The objective is to find the combination of glacial deposit hydraulic conductivity and recharge rate that best represents the observed water table. Modern ground-water software allows model calibration to be accomplished automatically after the user defines one or more objective functions.

The objective function is usually based upon the calculated difference between measured and predicted heads and/or measured and predicted stream flow. The VRML display allows the students to intuit the calibration directly, however. When hydraulic conductivity is too high, or recharge too low, the water table is far below the surface. Because there are perennial streams in the watershed, this clearly cannot be the correct solution. When hydraulic conductivity is too low, or recharge too high, the water table rises above the topography. This also cannot be a reasonable solution. If the student has a fundamental understanding of the interaction between surface and ground water, he or she should quickly arrive at the correct solution, wherein the water table lies just below the surface through most of the watershed.

After arriving at a satisfactory solution in this manner, the students can investigate the sensitivity of the model to hydraulic conductivity and recharge. Model error is displayed as a contour plot, where hydraulic conductivity and recharge are the coordinate axes. Students are asked to find the best combination of parametres to reduce the difference between measured and simulated heads. Based upon the shape of the error contours, they are asked to comment on the relative sensitivity of the model to hydraulic conductivity and recharge. If they could collect more field data, would the acquire a better estimate of hydraulic conductivity or recharge?

The VRML environment allows the student to see that the water table is a subdued reflection of the topography, and that it tends to be deeper under topographic rises. The comparison of topography and the water table illustrates a problem in ground-water modelling that is not usually noticed. In even the best-fit solution, the water table rises above the topography in limited areas.

This occurs because the model does not know the elevation of the land surface. Only hydraulic head values in the form of well levels, stream stage, or lake stage enter into the ground-water flow calculations. Thus, although the water table appears to reflect the topography, it is really dictated by the presence of internal boundaries. Students are asked how the model simulated water table would change if recharge and internal boundaries remained the same, but another 100 m of sediments were piled on top of the model (the answer is there would be no change).

Pedagogical Benefits

General principles of effective learning are incorporated into the ITD. Foremost, is the intuitive principle that learning-by-doing is the most effective method for teaching complex knowledge-domain-specific tasks. Students are given computer-based manipulatives that allow them to test their understanding of the relationship between geologic properties and ground-water flow.

We introduce these concepts in context of a larger system, to provide students a motivation for exploration. Visualisation-based learning can fail when students are disoriented and/or do not care about the larger problems. We offer these lessons in a laboratory environment that promotes cooperative problem solving.

It is difficult to compare our computer-assisted Ground-Water Modelling Lesson to "traditional" instructional methods, as there is no established approach to teaching ground-water modelling in general hydrology or hydrogeology courses. Popular hydrogeology textbooks usually approach the topic by describing the relevant finite-difference equations, and then discussing the construction of models and case studies. There are generally no hands-on exercises offered at this level. Hands-on training is typically offered in upper-level or graduate-level specialised, courses on ground-water modelling, which involve training in professional software packages.

The danger of introducing these software in lower-level courses is that, unless sufficient training and guidance is provided, students will become lost in the software and lose track of the main lesson. Audet and Abegg (1998) used professional Geographic Information System software in a classroom and found that novice users of the software tended to exercise lower-level cognitive skills than the more experienced users.

Without instructor supervision, students must resort to trial-and-error solutions to computer-based problems. Chang (2002) compared teacher-centred and student-centred computer assisted instruction of debris flows, and found that the self-paced group performed more poorly on a 30 item multiple choice test, possibly because they were overwhelmed with unstructured information.

The impact of using VRML in this lesson, rather than static images, is particularly difficult to quantify. In their recent review of visualisation in the geosciences, Libarkin and Brick (2002) note that rigorous assessment instruments specifically designed for visualisation of Earth systems are not currently available but are sorely needed. There do seem to be some obvious advantages to communicating three-dimensional spatial relationships in a VRML environment. It is very difficult, for instance, to compare the water level surface and topographic surface in two-dimensions.

It takes several minutes to become reasonably adept at handling the VRML object, but once the students become familiar with the interface they customise the view to their specific needs. This has the dual advantages of displaying the information in a manner best suited to individual learning needs, and involving the student in the learning process. It is interesting to note that some students rotate the landscape, whereas others prefer a map view. Some students "fly" through the landscape while others prefer to manipulate it as an object.

The number of 3-D visualisation tools for teaching geosciences appears to be increasing rapidly and VRML seems to be the present data format of choice. In spite of the activity in this area, the technology is by no means established. As is the case with many computer technologies, at about the time a VRML technology becomes stable it also becomes obsolete. This presents a difficulty to the development of classroom pedagogy that usually takes several years of field testing before it becomes effective. The Digital Library for Earth Science Education (DLESE) has begun to coordinate some of these efforts through meetings and online discussions and should result in a more concerted effort to MM 3-D content into the classroom. The value of VRML technology as a teaching tool for hydrology is unknown. As is the case with geoscience visualisation in general, better tools for evaluating the effectiveness of 3-D computer objects are needed. Expert evaluations in this area are certainly warranted given the significant effort being expended toward the development of visualisation and virtual reality pedagogy for the geosciences.

Encyclopedias comprise a major genre of printed and electronic reference material available in most libraries. The range and depth of subject content, and clarity of the writing style offered in major encyclopedias often impress librarians. Even so, encyclopedias are often underused, especially for instruction purposes. Patrons rarely ask for them by title and few instructors assign or even encourage their use by students. Many instructors discourage or even prohibit their use. However as the introduction to a comparative guide to encyclopedias notes, "Not everyone agrees with this reasoning ... Mindlessly insisting that students not use encyclopedias in the course of their school work is, frankly, a strange approach to education. It suggests a weak and rigid teacher who lacks understanding of the spirit of learning."

One of the reasons for this under use appears to be a common misconception by both students and instructors that encyclopedias are by definition general and simplistic. All too often, encyclopedic consultation is viewed as an academic weakness. Another reason is that most students and instructors are unaware of the existence of subject speciality encyclopedias that focus on a single subject field or even a few topics within a given field. Usually these sources are prepared for a sophisticated and specialised audience.

One of the purposes of this chapter is to suggest that the consultation of encyclopedias, especially the subject speciality titles, is often an academic strength not a weakness that can be encouraged by their inclusion in instructional programs. Since this chapter is written from a librarian's perspective, it is not suggestive of specific instructional techniques. One of the major functions of the library field is to create meaningful order of the literature of various subjects and to suggest meaningful patterns for the use of this material. Therefore a second purpose of this chapter is to provide a list of recently published geoscience related encyclopedias that are available in libraries throughout North America. The author infers that significant instructional potential for the secondary and undergraduate levels exists in these sources. It is up to the instructors to realise this potential by including these sources in their curricula.

It is relevant to note that this chapter is an outgrowth of the author's experience in teaching a one-credit university level course in library resources and research methods. The session on encyclopedias proved to be the most popular. Students would sometimes turn in assignment sheets with complementary comments scribbled in the margins. One particularly useful set of questions asked them to browse the library shelves and identify 9 different encyclopedias, 3 each in the humanities, social sciences, and sciences. The term "encyclopedia" has been used since the 17th century to designate publications summarising large segments of knowledge. Quite logically, the Encyclopaedia Britannien has one of the clearest descriptions of modern encyclopedias. It says:

An encyclopaedia is a self-contained reference work with two main aims: to include up-to-date knowledge about a particular discipline or group of disciplines and to make this knowledge conveniently accessible. An encyclopaedia differs from a dictionary in that it can explain subjects in detail, rather than merely supplying definitions of words and phrases. It differs from an almanac in that the information contained in the latter is dated. It differs from pedagogical texts in its attempt to be easy to consult and be readily understood by the layperson. To ensure comprehensiveness and depth of coverage, an encyclopaedia is generally written in the form of many separate articles, often by experts in the field. These articles include background and historical information as well as current material, which consists of varying combinations of text, tables, charts, illustrations, and, in the case of electronic encyclopaedias, audio and video recordings. Some encyclopaedias also offer study and learning guides, as well as yearly supplements that provide updates of various kinds.

Individual articles often contain bibliographies specifically designed to lead the reader to additional key material. The articles are edited forming a coherent work with a consistent style suitable for a particular audience.

Of particular interest to secondary and post-secondary educators, are the subject oriented encyclopedias that provide comprehensive summaries in broad categories such as science and technology, traditional academic subjects such as chemistry, topics such as sedimentation, or cross-disciplinary studies such as earth systems science. These may be single or multi-volume.

A leading authority on library reference material estimates there are 1,000 English language subject encyclopedias. This seems unusually conservative by other accounts. More importantly, both the quantity and quality of the subject encyclopedias exceeds that of the general encyclopedias.

5

Integration of New Methods

Mineralogy is of fundamental importance to the geosciences (solid earth, planetary, soil, hydrological, environmental, and ocean sciences) because the composition, structure, and physical properties of minerals ultimately control natural chemical and mechanical processes. Mineralogy has traditionally been one of the cornerstones of the geoscience curriculum, but recent trends in the job market, coupled with an explosion of new knowledge in all disciplines, have called into question the relevancy of the traditional curriculum. For example, Dr. Gordon Eaton, former Director of the U.S. Geological Survey, has been critical of the geoscience curriculum in a number of forums. Mineralogy in particular (along with the related disciplines of petrology and geochemistry) has been singled out as an example of arcane, irrelevant knowledge. His main point of contention is that this curriculum is more than 100 years old, and it has mainly been directed toward training of geologists for work in the extractive resource industries, whereas current employment trends require training in fields such as hydrology, education, engineering, and environmental geology. In light of demands to add new, practical components to the curriculum, hard decisions have to be made about which courses (or concepts, or methods) should be deleted or compressed to meet the contemporary needs of our students in their professional development.

However, as much as reforms are needed, one has to ask the question "Can any geoscience discipline be adequately explored or applied without a basic knowledge of the materials of the Earth?" We cannot hope to understand how the Earth and other planets work if we do not know what they are made of! Whether representing melting reactions near the Moho on a pressure-temperature diagram, or examining Eh-pH relations that control acid mine drainage while using minerals to immobilise hazardous waste, the relationships between minerals and their local environments have important implications for all geosciences and for society. Ignorance of mineralogy has cost our society dearly, as witnessed by the asbestos problem of the 1980's and, more recently, the ruling that quartz is a human carcinogen. Mineralogy is also particularly needed by K-12 educators: the first questions asked by students about geology are usually fundamentally grounded in mineralogy, such as "What mineral is this?" The problem is not that mineralogy is irrelevant, but rather that new methods and materials for teaching mineralogy are needed to demonstrate fundamental principles. We need to present this information in meaningful contexts to create a better learning environment. In this chapter, we address our own attempts at facing these challenges.

When we started teaching mineralogy, each of us taught the course the way we had learned it as students. The course is usually taught as a series of subjects (crystallography, crystal chemistry, classification, etc.) in a very linear fashion, starting with a set of supposedly simple material and progressing in a straight line to complex material. For example, a mineralogy course traditionally starts with crystallography. While it is incredibly important for many untold reasons, crystallography is very difficult to learn without being placed in some context. There is an incredibly large amount of vocabulary that goes with learning crystallography, and it is often the first time that students face college-level studies of abstract visualisation in three dimensions. After several weeks of crystallography (often without even the mention of a mineral name), the class moves on to crystal chemistry. Again, a large amount of material is introduced, completely distinct from what has come before. Real minerals are first mentioned about half-way through the semester, and the remaining studies of minerals and mineral classes typically march through the progression of systematic mineralogy beginning with elements, sulfides, oxides, and ending with silicates. Depending on the experiences of the instructor, these subjects may not be interrelated, but merely presented as a set of facts that together constitute the science of mineralogy. All too often, with time running short at the end of the semester, coverage of the silicates is limited.

After several years of teaching in exactly this sequence, it became apparent to us that this was not a good approach. Clearly portions of the course were improperly balanced. The majority of the minerals we asked our students to learn do not occur commonly as rock-forming minerals; e.g., a small number of silicates make up over 90% of the Earth's crust. Why not start with the really important silicates (i.e., quartz, feldspars, micas, amphiboles, and pyroxenes) the first day? Why are we devoting so much time to crystallography and crystal chemistry without relating them to each other (or perhaps more importantly, to geologic processes, and to the types of minerals that will occur in various parageneses)?

At this point (after 30 years of combined experience teaching this course), we stepped back and began to question what we really want students to learn from our one semester mineralogy course. We are grateful to the "Teaching Mineralogy" workshop in 1996, organised by John Brady and Dave Mogk, for bringing us together and forcing us to defend our vision of mineralogy to colleagues trained as petrologists. Subsequently, the first two authors of this chapter were fortunate to spend a semester jointly teaching a mineralogy course at the University of Idaho. Together we have arrived at the course goals that are discussed below. Given these basic goals, we then began working with JCD and MRLO to understand modern pedagogical theory, seeking ideas to improve how we can most effectively teach this material. We realised that learning comes from repetition of material moving from simple to complex, and from establishing connections between material as learning occurs. Thus we have incorporated several different teaching strategies into our courses, which are summarised here. We hope that this chapter will stimulate others to re-evaluate their own goals and methods for teaching this course.

The first step in developing a course is to set forth its goals. Most professors instinctively model their courses on the ones they took as undergraduates. For a number of reasons, this is probably not the best approach: we are not clones of our professors, each class has its own level of student intellectual ability, the material we are teaching is constantly changing, and newer research-based strategies on teaching science, as discussed below, show that these older methods do not work well. Accordingly, the following course objectives were developed. We believe this set should serve many, if not most, mineralogy courses currently being taught.

Introduce crystallography, crystal chemistry, and systematic mineralogy - Our goal is to help students attain a working knowledge of these basic concepts in mineralogy. What do these words mean? In this context, crystallography is simply the study of atomic arrangement. It is the science of how atoms arrange themselves to make crystals, and it has profound applications not just in mineralogy, but in chemistry, biology, physics, material science, and even mathematics. Crystal chemistry involves understanding the chemical make-up of those atomic arrangements. In other words, we want to know which kinds of atoms are where in the mineral, and why. Systematic mineralogy involves mineral classification and descriptions of minerals' physical properties.

Relate the physical properties of minerals to their crystal structures - Before students take mineralogy, they have probably thought about minerals from the perspective of hand samples; colour, hardness, streak, and other properties are taught in middle, junior and/ or high school as well as in most introductory geology courses. Thus, most students enter the course focused on physical properties, which in this context include any observable or measurable characteristics. We can use these physical properties by relating them to characteristics of mineral structure or chemical composition. To understand minerals is to use these interrelationships to our advantage, to help not only in mineral identification, but also in relating mineralogy to geology.

Introduce analytical methods used in modern mineralogy, especially the polarising light microscope - Ultimately in this course we must move away from hand sample diagnostic properties, which often yield incorrect mineral identifications, and into more unequivocal types of mineral descriptions. Analytical methods are an important part of mineralogy because they help identify and characterise minerals! If students have access to polarising light microscopes, they can learn one of the oldest and most useful tools in our field. It is also a useful skill to have when applying for jobs: for example, the entire asbestos abatement industry depends on people who know light microscopy! Depending on the accessibility of other analytical equipment, students can also have the opportunity to learn how modern mineralogists work.

Learn how minerals are classified and named - As in biology, paleontology, and other fields where hierarchies are important, there is a formal classification system for minerals. It is based on the kind of anion or anionic complex in a mineral's structure, and, to a lesser extent, on crystal structure. Sadly, a large percentage of geologists do not understand the difference between a mineral species and a mineral group, leading to constant (and unnecessary!) confusion in petrologic studies and even raising important legal issues because hazardous minerals are regulated by name. Mineral species names come from locations (where they are first found), appearance, chemical composition, and people's names. There are at least 4,205 officially Recognised mineral species names, and about 14,000 other mineral names that are in common usage. These include mineral groups, which are compositional variations on the same crystal structures, such as olivine, feldspar, and pyroxene. Through this course, we hope to give students familiarity with mineral nomenclature so they can use it correctly in their future lives as geoscientists, lawyers, medical professionals, etc.

Identify minerals in hand specimen and thin section, and with the aid of various analytical techniques - Mineral identification is a skill that is fundamental to many kinds of geology. By the end of this course, students should have the ability to identify many minerals with a hand lens on an outcrop, or in the lab with a thin section (a 0.3 mm piece of rock glued to a glass slide). The students' abilities to Recognise minerals will progress as they gain experience and have access to different analytical facilities!

Appreciate the influence of crystal chemistry on mineral assemblages and mineral weathering - This is a course about minerals, so we try to avoid talking about rocks whenever possible (that's the province of petrology!). However, sometimes the process of mineral identification is aided by knowledge of where the sample comes from. So we'll look into the ways in which mineral chemistry affects rock-forming and rock-breakdown processes, and vice versa.

Develop the ability to research and learn mineralogical topics individually and in groups -When we teach this class, we encourage our students to explore this field through a series of individual and group learning projects. We hope that they will be inspired to look beyond classical mineralogy, and to do some research of their own on some aspect of mineralogy that they find interesting. We also explicitly Recognise and do assessments based upon group and cooperative learning, and in doing so we remind our students that group collaboration is a necessary skill for employment in the 21st century. This approach incorporates the essential features of inquiry. Students are engaged by scientifically oriented questions. They give priority to evidence allowing them to develop and evaluate explanations that address these questions. Students must also evaluate their explanations in light of alternative explanations, and they must communicate and justify their explanations.

To accomplish the goals of learning mineralogy, we use four basic forms of newer types of pedagogy (i.e., learning methods and teaching styles) that should help students learn the material. These forms are: 1) spiral learning, 2) inquiry-based learning, 3) concept maps, and 4) interactive models and visualisation. Each of these will be discussed in detail below, with pertinent mineralogical examples. We have selected these methods because research shows that they are some of the best strategies for learning.

While these methods may be new to mineralogy, these are the methods that most of us actually use to learn in other contexts, and they are well documented as learning strategies. For example, spiral learning involves beginning with a new concept and then continuously reintroducing and reinforcing it as new, more advanced concepts are introduced. Consider a non-mineralogical example: when we first learned to cook, we learned the ingredients before we learned the recipes, but in making the recipes we relearn the characteristics of the ingredients. The same is true for mineralogy, where we first learn the ingredients (i.e., elements) and next learn how those ingredients are "mixed" together to make things (i.e., minerals). One step further in this analogy would take the different dishes (i.e., minerals) and combine them to make a meal (i.e., rocks), all the while considering the elements and minerals that make up the rocks. Continuing with the cooking analogy, inquiry-based learning is nothing more than questioning what might happen when we do something (substituting baking soda for baking powder in a recipe, for example), and then doing it to see the outcome (flat cookies!). For concept maps, the cooking analogy would concentrate on linkages between different types of foods, and interactive multimedia might be used to show the interrelationships in three dimensions. Incorporating these new methods into a mineralogy course becomes intuitive (and quite liberating!) once we dismiss our preconceptions (and previous experiences), and think about how we all learn!

Spiral learning - The idea of spiral curriculum was first developed by Bruner (who was in turn inspired by Piaget) as part of what is called "constructivist theory" starting in the 1960's, and it has been widely adapted in K-12 curricula around the world. It is based on the idea that learning is an active process in which students always construct new ideas and concepts based upon their current and previous knowledge. Its underlying tenet is that basic scientific concepts can be introduced to children in a form that is easily comprehensible to them at early stages of their education. The same concepts can then be revisited repeatedly at successively higher levels, enhancing and deepening students' understanding of the concepts, so that students are continually building upon what they have already learned. A spiral curriculum is a very powerful educational tool, as it enables educators to carefully stage their teaching of often quite complex concepts in a way that makes it intelligible and interesting to their students. However, constructive approaches are rarely used in college-level science courses, which are typically linear in the way they present material. Interestingly, professors are more familiar with this approach based upon their experiences with language classes, which are always taught this way.

There are a series of words along the spiral. These words start with the simplest components of speech at the bottom, and end with the more complex at the top. The first thing we must learn before we can learn any language is its alphabet. Once this is mastered, we start to make words from these letters, and then form sentences from these words. There are rules we use to make words (spelling), and there are also rules we use to build sentences (grammar). When we begin to compose sentences, we are using words, and in using words we must know the alphabet; thus one concept reintroduces and builds on the one below it. Lastly, we can put sentences together to form paragraphs. At each level, we are using the material from the level below, so we are reinforcing our knowledge of it.

We learn mathematics in a similar spiral fashion (in fact, did you ever notice how your algebra skills improved during calculus?). The first things we learn in mathematics are numbers; this is of course the "alphabet" of mathematics. The next whirl of the spiral is counting, which revisits the use of numbers, and the next logical step is addition. Once we can add, then we can subtract, but we cannot learn subtraction until we know addition. This process continues up the spiral. At each step, we reintroduce and reinforce the previous concepts. In fact, we often joke that when learning mathematics you never actually learn the last mathematics course you take because you only learn the previous course while taking the current one! The repetition of lower level concepts at each successively higher level of the learning spiral distinguishes this method from linear learning, in which concepts are introduced and then assumed to be understood from then on.

A similar theme can be seen in these two figures when compared to the previous figures on language and mathematics. In all cases we start with something very simple and build to more complex knowledge. However, in crystal chemistry and crystallography these two spirals are also linked to each other in a similar manner that the spirals in language and mathematics could be linked to each other (i.e., while one is learning a language one might also be learning mathematics). One cannot hope to communicate in mathematics if one only knows mathematics; one must also be able to speak a language to communicate the mathematical knowledge. In a similar manner crystallography and crystal chemistry are linked; we must understand both of them and their integration to understand either of them. This linkage of material is formalised in education theory based on a concept maps.

In our own courses, we also envision the interrelationships among concepts with a learning spiral. In the first week of classes, we introduce the "Top Ten Most Wanted Minerals", the six crystal systems, optical systems, and the "Top Eight Elements in the Earth's Crust." If we teach the course well, we hope that our students will remember this content ten years later! In the next 3-4 weeks we expand upon each of these topics, delving into crystal chemistry, crystallography, optical crystallography, and systematic mineralogy in the context of hand sample physical properties. The last 8-10 weeks of the course deal again with each of these topics in full detail. As we progress up the learning spiral for mineralogy, students continually reinforce the concepts mentioned previously, and they master simple subsets of material (e.g., 10 minerals and 6 crystal systems) before moving on to more involved material (e.g., > 150 minerals and 32 crystal classes). We find, of course, that we cannot cover material in as much depth as a traditional course progression; however, the material is covered in such a way that the students do, indeed, understand it better and retain it in the long term. We are in the process of formal assessment of the two methods of teaching to evaluate success in terms of students' achievements and knowledge gained.

Concept maps - Concept maps show how material can be linked in a nonlinear fashion. They have utility in showing students how concepts are interrelated, and they are also useful in helping instructors organise their courses. Concept maps, like spiral learning, have become very popular in modern learning theory, and have been applied to everything from foreign languages to law, and even the sciences. Novak (1991, 1995) provides a recent overview of concept maps.

Objects are placed within boxes, and a relationship is shown between these objects by using lines to connect them to other boxes. Geology is the central theme of this concept map. Geology is broken into six separate disciplines in this concept map, though of course many more could be included. One of these disciplines is mineralogy. Mineralogy is in turn broken into two separate subdisciplines: descriptive mineralogy and crystallography. Then the concept map shows how crystallography uses various techniques, X-rays and microscopes, and in turn how other disciplines such as chemistry, physics, and art, use the same instruments. Links are thus established between art and geology. This map is of course not comprehensive, because it could also show links between other subdisciplines in the field of geology and other disciplines outside the field of geology.

Another concept map shows how we link the goals of our mineralogy course. In this concept map minerals are the central theme. The upper four boxes show how minerals relate to other things we see in the world. First, minerals make up rocks. This is important for geology and is one of the main reasons students take mineralogy! Minerals also weather to form soils, which in turn affect agriculture. The other two boxes show that minerals make up bones and teeth, and minerals are also used in many important industrial applications. The lower portion of the concept map hopefully helps to more clearly define the goals for our course. It shows how minerals are classified, named, and identified.

The inquiry-based approach is the centrepiece of the National Science Education Standards: "Inquiry into authentic questions generated from student experiences is the central strategy for teaching science". We are not aware of a single course or textbook in the geological sciences that uses this approach, though it is being increasingly used in the social science and education literature. The need for this approach is nicely summarised by Alberts (2000) as follows, in a paragraph where the word "Mineralogy" could easily be substituted for "Biology":

> *"Where in a typical Biology 1 college course is the "science as inquiry" that is recommended for K-12 science classes in the National Science Education Standards? These courses generally attempt to cover all of biology in a single year, a task that becomes evermore impossible with every passing year, as the amount of new knowledge explodes. Yet old habits die hard, and most Biology 1 courses are still given as a fact-laden rush of lectures. These lectures leave no time for inquiry: they fail to provide students with any sense of what science is, or why science as a way of knowledge has been so successful in improving our understanding of the natural world."*

Alberts (2000) goes on to say that inquiry-based learning is far more efficient for long-term knowledge than the more traditional ways of simply being told how something works. In inquiry-based learning, students must be asking questions why and, in turn, trying to figure out the answers for themselves. Eventually students may be told the answers in class or read them in a book, but if they have already figured them out, or at least thought about them, they will remember them much longer, and it will make more sense to them than if someone simply told them.

We have not yet become comfortable enough with this concept to structure the entire course around it, but we use a combination of presentations of material that can be learned by experience, and material that can be learned through traditional methods. This accommodates students with different learning styles, as clearly stated by Welch et al.:

> *"Our stance is that all students should not be expected to attain competence in all inquiry-related outcomes, which science educators (including ourselves) have advocated in the past. For some students and in some school environments it may not be appropriate to expect any inquiry-related outcomes at all."*

As an example, a class on hand sample identification may begin by passing around a set of hand samples of different minerals. We then ask the students to think about a classification scheme that would allow them to organise the minerals into logical groups. Most students will instinctively group minerals by colour and, to a lesser degree, morphology. A list of descriptive criteria for organising the minerals is developed, and possible distinctions are discussed. Next, the chemical formulas for the minerals are written on the board, and students are asked to search for clues in the formulas to explain the differences in colour (almost always caused by the presence of iron in silicate minerals). They are then given the opportunity to reconsider their putative classification scheme in light of the new chemical information, and ultimately they usually come up with a rational scheme that is not too far from the conventional one. Eventually, we present the "Recognised" ways of identifying and classifying minerals, and end by summarising the appropriate terminology in this area (which is often not far from the descriptive terminology used by students). So, we start off by asking students to conceive of their own scheme, and conclude with the "official" scheme, making the transition from inductive to traditional learning. This is a very different approach from what we normally use in formal lecturing, it takes a lot of time and patience, and it is very different from the way the material was "taught" to us. It also requires that the professor have a thorough understanding of the material being covered, because it is sometimes necessary to defend the rationale behind whatever it is that you are teaching. We love rising to this challenge!

Interactive models and visualisation - One of the biggest problems in teaching mineralogy is the need to work "with inherently three-dimensionalcourse material, especially in a classroom with a two-dimensional blackboard. Most of us already rely extensively on visual aids including ball and stick models, coordination polyhedra, and optical indicatrices. However, unless the students interact with these materials, they are essentially static, and their teaching value is greatly diminished. The ideal is to pursue engagement with these materials, involving "student thought and interaction that goes beyond simple manipulation or movement via computer prompts".

Interactive tools are already used for many activities in existing mineralogy classes; several of these are described in this and previous issues of this journal and in the MSA workbook on this topic. The infamous wooden blocks used to teach symmetry are a good example of an exercise that helps students understand complex processes through direct manipulation. Many instructors employ styrofoam balls and wooden sticks to teach lessons about coordination polyhedra. Short of having students assemble their own crystal structure models (Gunter and Downs, 1991) visualising and interacting with mineral structure models is more difficult. For these and many other abstract or inherently three-dimensional concepts in mineralogy, computer animations may provide a means for directly interacting with course material.

Use of interactive video- and computer-based learning programs has clear advantages for students. Research shows that the use of 3-D models will allow students to discern patterns more quickly and to detect relationships between patterns or structures that are not obvious in 2-D. As noted in Bransford et al. (1999), "the ability of the human mind to quickly process and remember visual information suggests that concrete graphics and other visual representations of information can help people learn, as well as help scientists in their work." Interactive exercises have the potential to act as "tutors" to give students feedback on their understanding of the material. For example, the XTALDRAW program or CrystalMaker begins with a simple listing of symmetry and atomic coordinates and from these, displays animated drawings of mineral structures. If used in lectures, these animations are basically an extension of static images with the advantage of better illustration. A more engaging student activity is to let the students build or manipulate the input files themselves, so they can understand the effects or varying crystal class or atomic locations.

At the present time, there is a paucity of interactive computer-based tools for use in teaching mineralogy. Although there are many encyclopedia web sites available with information about minerals and mineralogy, none of them is interactive. The CD-ROM accompanying Klein (2001) and The Study of Minerals fall short of complex interactions with materials, although we have found the simple exercises contained in them to be very useful. Smith et al. (2000) have reported on an interactive CD-ROM for optical mineralogy. We are in the process of creating newer interactive exercises that require complex thought and engagement by students, but this task is challenging. We encourage colleagues to devote time to the development of interactive exercises for mineralogy; specifically needed are exercises that engage rather than simply illustrate!

Both MDD and MEG were fortunate to learn mineralogy the first time around from some wonderful instructors, despite the fact that "traditional" approaches were employed. However, the mineralogy courses of the 1960's and 1970's are still being regurgitated nearly verbatim in geology departments around the world, and this is, to us, a sad state of affairs. As both research mineralogists and teaching faculty, we feel obligated to do what we can to teach our mineralogy courses in the most effective way possible. Changing our courses to accommodate new pedagogies while sacrificing some "in depth coverage of material" has not been easy for us, but modern education research has convinced us that these changes are needed to make student learning more effective. We are hopeful that awareness of ongoing research into the effectiveness of new pedagogies, as well as development of improved teaching materials, will convince instructors who teach mineralogy to consider making some changes for the betterment of the discipline.

Geomicrobiology

Geomicrobiology is the study of the role that microbes play or have played in specific geological processes. Although Geomicrobiology studies has been around since 1887 with the initial discoveries by Winograsdky regarding H2S reduction by Beggiatoa, the field has received wider recognition in the last decade due to discoveries about the influence of microbial activities in shaping the habitable part of our planet. Stromatolites as the earliest geo-biological signatures and the search for life forms in Mars are among these important discoveries that are quite appealing to students. As a response to this trend, new and diverse interdisciplinary programs such as Interactive Geosciences

(University of Connecticut), Geology with Environmental Emphasis (Miami University), Geology and Geophysics (Louisiana State University), Geological and Atmospheric Sciences (Iowa State University) and Earth and Planetary Sciences (Washington University in St. Louis) have been established. All of these programs share a Geomicrobiology course to introduce main concepts of the field. While a Geology course is a prerequisite for the majority of the Geomicrobiology courses in Geology-related programs, this is not necessarily the case in Microbiology programs. The requirement of a Geology course to teach Geomicrobiology imposes an important curricular constraint to non-geology majors that are interested in this emerging field. To overcome this barrier at the University of Puerto Rico, Humacao (UPRH) Campus, we have taken advantage of the Special Topics Course to offer the first Introductory Geomicrobiology course in Puerto Rico. The design and implementation of the course was supported through an NSF Research in Undergraduate Institution (RUI) in Geomicrobiology.

Research Description of the at Undergraduate Institution (RUI) in Geomicrobiology

The RUI in Geomicrobiology is an educational initiative that is part of a Microbial Observatory (MO) proposal established at the Cabo Rojo salterns in Cabo Rojo, Puerto Rico. The Cabo Rojo Salterns Microbial Observatory (CRSMO) is dedicated to identifying the microorganisms present in the crystallising ponds and microbial mats that prevail at the location. Beyond the mere identification of the organisms, we study their possible role in the geological transformations that take place at the locations with special interest in the microbial mats communities. Due to their coloured layers and the sticky nature of the top layers, these microbial mats are quite appealing to our undergraduate students from different backgrounds (i.e., Geology, Microbiology, Marine Biology and Biology). By dissecting the mats the students are able to visualise the direct contact of microorganisms and the minerals associated to them. We have recently reported the use of the microbial mats as tools for promoting active and effective learning of basics concepts in Geomicrobiology. In this chapter the students attended a series of workshops and described the following aspects as the most important outcomes of their participation in the RUI program: (a) the opportunity to study and learn about new and different sciences disciplines, (b) the use of microbial mats to integrate different science disciplines, and (c) the capacity to work in multidisciplinary teams to learn from their peers' disciplines' backgrounds. In addition, to conduct research projects and attend several workshops during the academic year, students from the RUI program also participated in summer experiences at the Marine Biology and Integrative Geosciences Departments from the University of Connecticut (UConn) in Avery Point. In the summers of 2003 and 2007 students have attended intensive Geomicrobiology workshops consisting of a week of lectures and experimental work.

The experiments conducted included microscopic analysis of microbial mats, setting up enrichments for sulfate reducers and methanogenic organisms from the anoxic layers of the mats, in situ gas measurements and determination of oxygen profiles from sediments around UConn. At the end of the workshop the students were responsible for presenting their findings in oral presentations. Further information on the CRSMO research and educational initiatives can be found via the internet at the web page of the RUI program.

As mentioned, to complement these educational efforts we offered a Special Topics Course in Geomicrobiology to undergraduate students from the UPRH. The course was designed to introduce students to the field of Geomicrobiology using eight main topics ranging from role of microbes in Early Earth to the search of possible extraterrestrial life. In this work, we describe the objectives, structure and assessment of this new Introductory Geomicrobiology Course specifically designed for non-geology majors.

A total of 13 undergraduate students enrolled in the Introductory Geomicrobiology course. Forty-six percent of the students attending were majoring in Biology, 38% in Microbiology and 15% in Marine Biology. Interestingly, only one of the students had taken Geology or a Geology related course before (Coastal Geomorphology) and four had never taken a Microbiology course. Since no prerequisite courses were required for taking the course, the student population consisted of sophomores, juniors and seniors.

Main Concepts and Learning Goals of the Course

The Introductory Geomicrobiology Course was designed to give students a broad overview of major topics or concepts in the field. These topics were reduced to:

- Role of microbes in geological transformations of our planet
- Early Earth ecosystems and the importance of microorganism in the establishment of life on our planet
- Microbial diversity of ecosystems that simulate Early Earth development

- Search for geological signatures in our planet
- Search for biological signatures in other planets
- Regulations regarding possible contamination with extra terrestrial life forms
- Survival and resistance of life forms in extra terrestrial simulated environments
- Conditions necessarily for extra terrestrial life.

The main learning goal of the course was to read, present and discuss recent research investigations that cover these eight topics, so as to provide the students basic knowledge about the role microorganisms exerts over geological transformations in our and other planets. This goal was met through an active learning environment where students discussed some of the major geological transformation of microorganisms (i.e., carbonate formation and dissolution) as well as some of the methodologies currently used to detect extraterrestrial life forms. An example of the latter was a detailed explanation of how X-Ray diffraction analysis help discern the biogenicity of certain minerals by their patterns.

The structure of the course was essential to help students understand these main concepts. On the first day the PI provided a lecture about Geomicrobiology and its important implications in the Early Earth development. As four of our students have never taken a Microbiology course before, in this lecture we exposed our students to general Microbiology concepts such as the sterility and aseptic techniques required for the growth of microorganisms and requirements of media to support growth.

In subsequent classes students began the discussion of review papers such as Erlich, 1999 to explain basic concepts of Microbiology and the different mineralisation reactions that take place in the microbial world, with special emphasis on the diversity of microbial forms that are responsible for them. To help students become familiarised with microbial diversity analyses we then selected one publication that relates microorganisms and their geological transformations. This publication was selected because it presents basic and general information, it is easy to comprehend for students of all backgrounds and it was conducted in an ecosystem that they are quite familiar with because it is located within Puerto Rico. The paper was useful to describe some molecular tools needed to analyse the community structure of an ecosystem. For example, we discussed techniques for DNA extraction, PCR amplifications, construction of genomic libraries and the use of phylogenetic analysis to determine community structure. Once the students were familiar with microbiology and molecular concepts of the papers discussed we then concentrated on the discussion of more complicated scientific journals with techniques such as X-ray diffraction to analyse minerals, differences among minerals from biological origins and the setup needed for experiments that simulate extraterrestrial conditions.

The structure of the course centred around a Journal Club format in which each student was assigned a research publication. Although the selection of articles was voluntary, each publication was restricted to the eight major Geomicrobiology topics and concepts previously discussed. The first day of the course the students received an introductory lecture about topics such as; What is Geomicrobiology? and Which are the most important biogeological transformations on our planet? The purpose of this lecture was to make sure that all students starting the course had similar knowledge about Geomicrobiology, independently of their background or year of study. At this initial lecture we also explained to the students the format to be followed in their oral presentations and discussed the main criteria to be used for their evaluations in the presentation and the course. The course evaluation was described in a syllabus given to the students on the first day of classes. The overall evaluation of the course was obtained from the presentation, with a 50% value, attendance, with a 10% value, and participation was worth 40% of the grade.

Educational Strategies

To promote an active learning environment during the Geomicrobiology course we used two main strategies: oral presentations of a recent publication in the subject followed by a Gallery Run activity or a short examination (quiz). We used quizzes to compare the student learning acquired independently by each student with the knowledge acquired during the Gallery Run. After each activity we conducted a period of discussion to clarify any remaining doubts.

Structure of Oral Presentations

Each student was responsible for presenting an assigned scientific journal article to the rest of the class in an oral presentation supplemented with PowerPoint files. Each presentation consisted of the basic elements of a common scientific article such as introduction, methodologies, results, discussion and major conclusions.

To help the students become familiarised with the main authors that wrote the publication, we requested that they looked for a picture of the facilities where the research took place as well as personal aspects of the researchers such as prior education and hobbies. Consequently, as part of their introduction, each student had to mention something particular about the first author or group leader of the publication.

They also were asked to provide background in the main topic to be covered in their presentation. Some of the students provided information about the characteristics of the Early Earth and atmospheric conditions at the stratosphere. Others answered questions such as; Which are the most resistant life forms on Earth? What is the difference between a mineral and a rock?

Another unique feature of the presentation schedule is that in the methodologies section each student was responsible for describing the basic operation of each one of the instruments needed for the study. For example, students needed to describe how scientists were able to mimic conditions in the stratosphere using a combination of gamma and UV radiation; other students described the main characteristics of the materials used for the insulation modules used for Caenorhabditis elegans outer space trips.

No special requests were imposed in the sections regarding the discussion of results, discussions of such findings and the presentation of the main conclusions of the work. Time for the oral presentation was limited to 20 minutes for each student. At the end of each presentation, the students had up to ten minutes to ask questions regarding the journal discussed. To encourage student participation a bonus point was added to each student that articulated a question in the discussion period. Indeed, a fundamental component of the course was the class discussions held after each oral report. We encouraged discussion in informal and formal ways. Informally, students were stimulated to ask their own questions about the material presented by their peers. Formally, they participated in a class discussion technique called Gallery Run, which is a sped-up version of the Gallery Walk. In a traditional Gallery Walk students have about 5 minutes to answer each question, but in the Gallery Run student only have a maximum of three minutes at each station.

Description of Gallery Run Activities

In Gallery Run, students from different backgrounds (Biology, Microbiology and Marine Biology) were arranged into teams of three or four. Each team was provided with a different coloured marker to write the answers to questions about the presentations on one piece of paper. Before the activity, the roles of a leader, recorder and reporter were assigned to group members.

The leader encouraged participation and kept the group focused on their main task, the recorder wrote group responses on paper sheets and prepared the final answer written report, and the reporter presented a summary of the group's thoughts on a question to the class. The roles were alternated between each team member throughout the course. To ensure that all students played all three roles, cards with an illustration alluding to the role were prepared and hung around each student's neck. On the back of each card the student wrote her/his name and the date she or he performed each specific role. Each week the professors in charge of the course verify that the students were indeed conducting a different role. The professors were also responsible for writing the questions based on the publication discussed that day, and ensured that the questions addressed higher-level thinking skills according to Bloom's Taxonomy. Once the reporter presented their summary to the question asked, we had another period of discussion in which all ideas and concepts were discussed and doubts were clarified.

Description of Quizzes

Short exams or quizzes were randomly given after some oral presentations. Quizzes consisted of a similar number of questions as those utilised in Gallery Runs; however, each student had to answer them individually. The time allowed for answering the questions was the same and they were written according to the Bloom's Taxonomy as were those used in the Gallery Run activity, Similar to Gallery Run activities, once the students answered the questions, they were discussed in order to clarify any concept or idea that they had not understood.

An important aspect of the course was to determine the extent to which students met the learning goals, as well receiving their feedback regarding the activities implemented. Consequently, we developed specific strategies based on both formative and summative assessment. Formative assessment strategies provide an immediate evidence for student learning and give the opportunity to improve it. Summative assessment, on the other hand, evaluates a particular activity to determine student's skills, knowledge or effectiveness of the activity itself.

Formative Assessment

Formative assessment included all the class discussion techniques, including the two sessions of questions asked by the students after their presentations and after the Gallery Run/Quizzes. Such sessions help the students reinforce and/ or clarify any doubts they had concerning the article presented that day in class. Furthermore, we used quizzes as an additional instrument to detect student understanding of the material presented in class. Three criteria were used to assess the quality of student written answers, both in Gallery Run and in quizzes: concept understanding, use of scientific language and clarity of thought.

Summative assessment included the scoring rubric designed for the oral presentations and a Gallery Run questionnaire to measure student achievement and instructional effectiveness. As mentioned before, students' understanding as well as their ability to communicate science effectively in their oral presentations was assessed. Oral reports were ranked from exceptional, good, satisfactory, to needs improvement, on a scale from 4 to 1.

The Gallery Run activity was formally evaluated by the students at the end of the course using a thirteen question survey previously described for Gallery Run activities. The survey was anonymous, non-graded and consisted of thirteen questions. Only two questions were open questions (question 1 and 2) and the rest were general observations to be evaluated by the student using a rubric (from question 3 to 13). Questions seven and eight allowed enough space for the student to explain her/his answers.

The main learning goals of the Introductory Geomicrobiology course for non-geology majors were: (1) for students to present and discuss research studies about the role of microorganisms in the geological transformations of our planet as well as (2) understand the adaptations required for the establishment of extraterrestrial life. To meet both goals we used a series of educational and assessments strategies and in this section we present our major findings.

1. Oral presentations in a Journal Club format were an effective way to introduce non-geology majors to tlte Geomicrobiology field. All the students enrolled in the course conducted an oral presentation on an assigned publication following a Journal Club format. Each student was evaluated utilising eight main criteria. Oral reports rubrics' global average scores were 3.20 (80%), with much of the evaluation relying between good and

exceptional. The highest scores (> 3.20) were reported in student organisation, authence interaction, questions and answers and maintenance of eye contact during the oral presentations. Students' explanations of concepts and theories were accurate with few errors in information, meaning that the students were able to read, understand and communicate data regarding introductory concepts in Geomicrobiology.

2. Student performance was stronger in groups than individually: Gallery Run versus Quizzes. Concept understanding, use of scientific language and clarity of thought were stronger in the answers given by the students as a team in the Gallery Run activity than those given individually in quizzes. Total score value for all written questions in Gallery Run was 103.73, while a total of 85.07 was obtained for individually answered questions. Since all questions were always answered in the Gallery Run activity but not on quizzes, when computing these total points, the two Gallery Run synthesis questions were omitted since this type of question was not asked on quizzes.
3. Gallery Runs were more effective in promoting class participation than individual sessions of questions. To increase student participation we offered a bonus point for each question formulated by students after each oral presentation. Only seven students in the course asked questions and each student had an average number of 3.5 questions throughout the course. Questions were basically of two types: those in which students wished to reinforce their understanding about a particular graph or table presented; and those with which they sought a deeper explanation of a concept or theory. Even though a bonus point was granted for each question made, half of the class never asked a question. This contrasts with student behaviour reported during Gallery Runs, where students were eager to talk and discuss the material.
4. Gallery Runs were the preferred class discussion technique used in the course.

A questionnaire was used to asses the relevance of Gallery Runs as a discussion technique in the course. All students agreed that the technique was easy to use, that the instructions for participating in the Gallery Run activity were clear, and that it really encouraged collaboration among the students (more than with other class discussion techniques). They also commented that through Gallery Run discussions they gained a more complete understanding of the topics presented.

Interestingly, when the students compared Gallery Runs to quizzes, they highlighted that the technique provided a deeper understanding of the journal assigned in class due to the opportunity to work collaboratively, learn from their peers and share ideas. When they answered questions individually their understanding was not as complete. One student mentioned on the questionnaire that only when quizzes were discussed by the professor during the question session was he able to understand the material to the degree he did with Gallery Run activities.

Summarising other answers from the questionnaire, all students (100%) indicated that they would like to participate in another course utilising Gallery Run activities in the future. The Gallery Run activity was favoured by the students due to its multiple advantages such as its ease of use, its quite dynamic nature, its allowing for collaborative' work, and that it provides an opportunity for more complete discussion as well as a deeper understanding of the publication assigned.

Eighty-five percent of the students agreed that the wording of Gallery Run questions was clear, while 15% was neutral (question 9). Regarding the timing for discussing each topic at the learning stations (question 10), 69% agreed that they had enough time to discuss each topic at learning stations but 23% was neutral and 8% disagreed.

We have described the objectives, structure and assessment of a new undergraduate course designed to introduce non-geology majors into the field of Geomicrobiology. One of the achievements of the course was to expose students to major topics and concepts in Geomicrobiology throughout different activities which were far from the traditional lecturing teaching style. Because we were aware that none of our students have a background in Geology or Geochemistry, we needed to make sure that they were not only exposed to the new information but also that had the opportunity to process it. To engage students and enhance their learning experience we promoted, from day one, an active learning environment.

The first way to actively involve the students in the course was for them to present oral reports on a current scientific journal article about Geomicrobiology. The oral presentations encouraged students to read, present and discuss recent Geomicrobiology research regarding eight major concepts in the field. Our results showed tli it the students acquired the necessarily skills and basic knowledge for presenting an oral report in the Geomicrobiology science field, our main learning goal. The fact that our students' oral reports rubric scores ranged between good and exceptional is the best indicator. Students felt quite comfortable with their presentations, were able to maintain eye contact and answered most of the questions. Nevertheless, in some instances we observed that students lack the knowledge to explain molecular techniques that are relevant to the field. For example, in the presentation of the article Community Structure, Geochemical Characteristics and Mineralogy of a Hypersaline Microbial Mat, Cabo Rojo, PR, the student failed to include an explanation of the T-RLFP technique used to assess changes in the microbial community structure in their presentation, as well as not mentioning the mineral composition of the mats. Fortunately, the discussion of the articles by the professors at the end of the course provided an opportunity to supplement information omitted in students' presentations and to correct such difficulties.

Another way we promoted active learning during the course was to encourage both nonstructured and structured discussions. According to Bonwell and Eison (1991, in-class discussion is one of the most effective strategies for promoting active learning. Indeed, the assessment of our structured discussions using the Gallery Run technique revealed superior student participation and improved the quality and quantity of their answers to written questions when compared to quizzes. Student answers to written questions provided us with direct evidence of their ability to understand information about Geomicrobiology through three different criteria: concept understanding, use of scientific language and clarity of thought. As previously reported students did better answering questions when they work in teams rather than individually, resulting in more complete and coherent answers.

This was also true regardless the cognitive level that the question involved (knowledge, comprehension, application, analysis or synthesis). For example, when an analysis-type question on a quiz asked students to make inferences about the considerations for successfully conducting an experiment based on the discussion of a journal article describing how Caenorhabditis elegans survived atmospheric breakup of STS- 107, Space Shuttle Columbia, students showed difficulty arriving at a logical answer individually. Similar types of experimental approaches were questioned when discussing prior journals during Gallery Runs and the students had no problem in responding successfully. Indeed an improvement was observed when students answered analysis type of questions as groups. We have previously determined that students working in groups in Gallery Runs posses a more complete understanding of the material as a direct result of the active discussions of the students on the team than when they worked on their own. Student's evaluations obtained through the questionnaire and the evaluation of the course activities concur with such findings.

The lack of an educational setting in other courses where higher-order thinking tasks are routinely assigned and assessed to encourage cognitive development could be one of the reasons why students present difficulty when dealing with these types of questions on their own. Since discussion is superior to lecture for promoting higher-order thinking skills involving analysis, evaluation and synthesis, a discussion technique such as Gallery Run could be implemented as a regular activity in other science courses to help students develop such skills.

Another advantage of using Gallery Runs is the multiplicity of roles (leaders, reporters and/or recorders) in which the students are able to develop both oral and written skills. As students needed to participate as reporters in each group at least one time, they develop skills such as correct technical writing of relevant scientific parametres (i.e., units for total radiation, pressure, how to report cell death during a mission or total bacterial counts after the vessel returns). In the role of speakers, even the more introverted students were encouraged to participate, a task that helps their self-confidence increase. One of the most important outcomes of the Gallery Runs was that the students were able to work as multidisciplinary teams. Even though students were from different backgrounds (Biology, Marine Science and Microbiology) they were able to effectively communicate and share knowledge from each field in order to answer the questions. The capacity to work on multidisciplinary teams is very important in Geomicrobiology, where teamwork of professionals from different backgrounds is virtually mandatory.

As a result of our assessment, we can conclude that most of our non-geology major students learned basic Geomicrobiology concepts even though our University does not possess a Geology Course in its curricula. Such a limitation can then be overcome if an active learning environment in the classroom is implemented using oral reports and discussions as main class activities. As previously reported, class discussions not only contributed to keep the excitement in the class but also promoted scientific literacy, which was an important goal we needed to accomplish. Our course provided an effective environment for learning Geomicrobiology, especially for non-geology majors with very little background in the subject.

Teaching Metastability

A now-common element of undergraduate science curricula is an introduction to reading the peer-reviewed primary literature. There are a number of compelling reasons for having undergraduates read journal articles in addition to standard texts, an important one being that those students who continue in science will need to be comfortable with consulting primary sources. Reading scientific papers from the literature also brings students closer to the goals and process of scientific research, which is important to students preparing for careers in science and to non-science majors. The expectation of clear scientific prose from students also necessitates a solid familiarity with scientific writing, including primary sources. In particular, Geoscience students need a familiarity with the primary literature, because ca. 70% of prospective graduate students in the Geosciences select graduate institutions based on their future advisor and probable research.

There are a number of possible approaches for exposing students to journal articles. Many times students will have to navigate the primary literature for the first time in the context of a term paper assignment. Although the term paper project itself may be highly structured (with an approved project, outline, annotated bibliography, etc.), the actual foray into the literature is often unique to the student and their research topic. As a result, the student's difficulty with the selected papers is a function of the student's familiarity with the author's methodology, the sophistication of the reasoning used, the clarity of the writing, and the student's skill in reading primary sources.

Some authors suggest a more explicit introduction to the primary literature for undergraduates. The form of this introduction depends on the level of the student and the goals of the exercise, and can include interpretation of landmark papers in the discipline, group-led exploration activities, or structured in-class discussion. Some other interesting approaches include the comparison of papers from the primary and secondary science literatures describing the same research results, the writing of abstracts for 'decapitated' research papers, and a 'mock-review' of a scientific paper. Often the general guidelines for reading scientific papers given to students have a format similar to the paper itself; concentrating on objectives, methods, interpretation, and hypothesis-testing. In this chapter I outline an approach that I have found successful in introductory and intermediate geoscience classes: a guided reading. Briefly, this method asks students to respond in writing to a number of questions unique to the journal article being read. This approach is also applicable to book chapters and some government geologic reports. The guided reading is distinct from simply assigning a series of factual questions to answer while reading the article. Instead, the questions are designed to step the student through the approach and logic used in the paper requiring progressively more sophisticated understanding and reflection on the methods, assumptions, and wider implications of the research. This method is meant to supplement other approaches used in the Geosciences to introduce undergraduates to the literature, and is especially appropriate in situations where students are in the early stages of introduction to science or a particular geoscience sub-discipline. The goal of the guided approach is to teach good habits for reading scientific articles, and to provide an entrance-point into the article for students who have trouble knowing where to begin. A guided reading assignment is also useful to set the stage for subsequent in-class discussions or debates. If assigned beforehand the questions provide a useful base-line of knowledge that can be used by the instructor while facilitating in-class discussion.

6

Building Bridges between Field and Laboratory

The National Science Education Standards (NSES) (National Research Council, 1996) advocate to learning "science as process". Inquiry-based learning, where students are engaged in observation, inference and experimentation, is a central theme. Numerous authors describe the benefits of constructivism in enhancing student learning in the sciences. Constructivist activities are not the traditional laboratory exercises that tend toward cookbook and confirmatory style problems. In contrast, students are engaged in open-ended activities that promote challenging their prior knowledge, encourage peer teaching and provide more opportunities for instructors to evaluate student learning. These concepts are useful in developing project-based laboratory activities that encourage hands-on participation in doing science.

Hands-on laboratory exercises are effective learning tools in numerous settings. In hydrogeology laboratories, several authors demonstrate the effectiveness of conducting experiments to determine the physical and hydraulic properties relating to groundwater flow. Nicholl and Scott (2000) introduce a simple and inexpensive permeametre for students to explore Darcy's law. In their laboratory activities, students investigate the effects of various parametres such as particle size and fluid viscosity on hydraulic conductivity. The authors also present suggestions for additional studies possible with the described apparatus. In this chapter, numerous hands-on laboratory activities are incorporated into the integrated lab-field project, and several methodologies and experiments suggested by Nicholl and Scott (2000) are further developed and adapted. Numerous authors demonstrate that field-based research experiences can be effective learning tools, attract students to the sciences, and make science more relevant. Huntoon et al. (2001) conducted a summer field course aimed at undergraduates and K-12 teachers. In their study, they attempted to quantify the effectiveness of the field research experience. Despite a small sample size, the authors suggest that assessments of lower- and higher-order cognitive skills showed immediate results among participants. Field experiences are also an important component for understanding hydrogeological concepts. Fletcher (1994) suggests that field exercises are a critical component to most geoscience courses, including hydrogeology Therefore, development of a field laboratory facility becomes essential for teaching undergraduates in this area.

Project-based exercises, extending beyond a single three-hour lab session, are effective learning methods as students take ownership of their projects. Soreghan and Soreghan (1999) describe a multi-week lab sequence in which students complete field and laboratory work. The ultimate goal of their project is the use and application of a variety of data sets to describe the structure and geologic history of a sedimentary basin. McKay and Kammer (1999) have integrated hydrogeology exercises into a mapping-based field camp. The goal of their hydrogeology component is to introduce basic concepts of groundwater flow, demonstrate the influence of geologic variability on flow, and introduce common field methods.

Students use and introduce common the lab to determine hydraulic conductivity of the material, and couple this determine hydraulic exercises in water table mapping and aquifer pumping tests. Field observations and data collection are conducted in a sand and gravel quarry. These data are used in later labs to construct and test small-scale aquifer models. Assessment of student learning using pre- and post-course interviews, final exam questions and student surveys indicated that students enjoyed the field-lab mini-course, felt more confident in their own scientific abilities and showed that students were able to understand the concepts covered in the mini-course better that those utilising other methods. Undergraduate geology and water resources majors are required to take a course in groundwater hydrogeology during their junior or senior year. Additionally, this same course serves as an elective for general earth science and environmental science majors preparing for secondary education or various environmental careers. By the time they take this course, students complete physical geology, two semesters of chemistry, and at least one semester of calculus, physics and applied statistics. Additionally, they typically complete several upper division courses in their majors as well. The earth science majors tend to be less well prepared as their curriculum requires substantial breadth in all the sciences, but not as many upper division credits in the geosciences.

The course is offered in a traditional lecture-laboratory schedule with three 1 hour lectures and one 3 hour laboratory per week. It focuses on the fundamental aspects of the hydrologic cycle with emphasis on the subsurface. Relationships between the physics of fluid flow and the physical properties of rocks, sediments and soils are emphasized to provide a basis for examination of the management and environmental aspects of groundwater hydrogeology. Stated learning outcomes for the course include:

- Define the major components of the hydrologic cycle
- Identify the principle components of subsurface hydrogeologic systems
- Formulate conceptual models that describe the hydrogeologic system under study
- Solve common expressions and equations that elucidate the physical and hydraulic properties of groundwater systems
- Design experiments that test a specific hypothesis
- Evaluate the quality of experimental data in both a qualitative and quantitative manner.

The laboratory portion of the course has evolved over the past four years from a series of "paper" labs to the current format of a dominantly project-based lab sequence. This project-based approach offers students more hands-on experiences, and better addresses the stated learning outcomes of the course.

The laboratory portion of the course is composed of two project-based exercises, 1) physical and hydraulic properties, and 2) groundwater chemistry and contamination. Although this does not represent a novel approach to laboratory design, the lab outline presented here for the first project attempts to incorporate multiple aspects of both hydrogeology and related subject areas such as sedimentary geology and soil science.

The first project-based laboratory experience, lasting nine weeks, involves the analysis of the physical setting and hydrologic properties of a potential aquifer. The aquifer is represented by an unconsolidated sandy deposit exposed in a local sand and gravel quarry. This glacial deposit is an esker of Wisconsin age. Although the sediments are not saturated, the ease of access and the ability to make observations and collect samples in three-dimensions offers many advantages. This physical setting also provides the opportunity to develop links to other geosciences courses that the students have taken or will be taking prior to graduation. This laboratory project is divided into nine week-long sessions. Each student is responsible for producing a final lab report discussing the results of the entire project. I encourage students to submit draft summaries of methods and results for each week's lab to keep them on schedule.

Students will construct permeametres during week five. Permeametres are easily constructed using common plumbing supplies. Each group is given a set of partially assembled parts. Prior assembly is necessary, as some parts need to be sealed to prevent leaks. Permeametres constructed for this series of labs cost approximately $65 per unit. The main body of each permeametre is constructed from 10.08 cm (4 in), schedule 40 PVC pipe. The permeametre body is 50 cm long with an I.D. of 10.08 cm, giving a cross-sectional area of 79.8 cm^2 and total volume of 3990 cm^3. Headers are constructed from common plumbing supplies found at the local home centre with screen installed to support filter paper. The bottom header is sealed at the base, and course gravel is added below the support screen to reduce the volume of the header. Water is introduced into the sand column through the bottom header and flows upward to avoid gravity drainage problems. Water is added through large diametre clear tubing from a reservoir. A large funnel situated above the top header is usable as a reservoir, but continual filling with small volumes of water is necessary to maintain a constant head during testing. The top header is left open and acts as an overflow chamber. This allows for easy determination of the water level exiting the permeametre. Large diametre clear tubing is used to allow easy discharge of water.

During the first week of the sequence, we discuss the overall goal for the laboratory (integration of field observations with laboratory data). I present guidelines for completing lab exercises, including time frame for each lab, group versus individual assignments and responsibilities, reporting requirements and background information. The students then have the opportunity to work with a complete data set to evaluate the nature of spatial variability and consequences of subsurface heterogeneity in a real unconsolidated water table aquifer.

The second week of the sequence is spent in the field making observations and collecting samples for further analyses. A local sand quarry serves to provide easy access to subsurface material. Using the sand quarry allows students to see in three dimensions the natural variations in the subsurface that would otherwise be hidden if using only core samples from an unconfined aquifer. A section of the quarry is selected to provide pit walls at an approximately 90 degree angle to each other. Using tape and compass, students work as a group to map the orientation and extent of the exposure to be studied. The overall dimensions of the sampled area are approximately 10 m by 20 in, but may vary at the instructor's discretion or as field conditions warrant. The students also produce scaled field sketches of the exposed cut face taking care to note distinctive bedding features in the unconsolidated material.

Samples are collected at random intervals in X and Y space with each student being assigned a location. If approaching a 90 degree angle, the walls of the cut face may become the X and Y axis, or a north-south and east-west orientation may be used. A corner of the sampling area is designated as the origin for the purposes of measuring distances in the X and Y directions. The Z component is measured from an arbitrary datum established at the time of sampling. The easiest, and most relevant is the surface of the upper bench. Using a level string line, the assumption that the ground surface is a legitimate datum may be tested. If the students determine that the assumption is not valid within an acceptable margin of error, a method of correction must be established. What is acceptable is discussed in the field in the context of the methods to be used to measure and collect samples. Using the upper bench as Z equals zero further familiarises students with measuring units of depth below ground surface, which is commonly used in hydrogeologic studies. Samples are collected from three depth intervals, giving each student three samples for further analysis.

Discrete samples are collected using a hand auger or thin walled sample tubes. Hand augers are used to collect samples from designated sampling points not along the cut face. Augers are advanced, and material discarded until the bucket is centred at the specified sampling depth. Samples are extracted from the auger, placed in plastic bags, and lebelled. Small aluminium cans or sections of 5 cm (2 in) PVC pipe serve well as a sampling tool for those samples collected from the cut face. The volume of the cans or pipe sections is measured before fieldwork with a volume of approximately 200 cm^3 being ideal. The sampling tool is driven into the pit face, centred on the measured depth interval. This may be accomplished by hand, or by placing a flat board on the back of the can or pipe and gently striking it with a rubber mallet. Once the can has been fully driven into it face, it is carefully excavated. A knife or spatula may be used to trim excess material from the ends. The sample is removed from the sampling tool, placed in a zip lock bag and lebelled. Upon our return to the lab, all samples are weighed and placed in a 105 deg C oven for at least 24 hours to dry.

A bulk sample should also be collected at this time for use in specific retention and permeametre studies to be completed in weeks 4 and 5, respectively. The bulk sample should be collected from the same location as one of the discrete samples. This is easily accomplished from the cut face. Analyses in the following weeks will provide specific physical property information regarding this sample.

Particle size distribution is determined for all samples. This may be the most time consuming step in this week's lab, depending on the equipment available. For this lab, students use a Beckman-Coulter LS 230 laser diffraction particle size analyser. This instrument uses the diffraction of light to determine the size distribution of particles in a streaming slurry. The instrument is easy for students to use with minimal instruction and supervision, and allows rapid analysis (-5 min/sample) of their samples.

Additionally, the instrument provides good graphics and data files that are easily transferred to spreadsheet software for further analysis. If sieving is used, it is justifiable to limit this to dry sieving at one-half or one-quarter phi intervals, ignoring the separation of silt and clay. Students will have the next few weeks to analyse their particle size distribution data. Graphs of particle size versus percent passing are constructed with particle size described in millimetres and phi. From these graphs, students calculate various statistical parametres such as uniformity coefficient, graphic mean (M_Z, inclusive graphic skewness and graphic kurtosis. Students will then combine the entire class data in one data set for graphical analysis of the spatial distribution of these parametres.

Students work in groups to build a device to determine specific yield and retention values for the bulk samples. The simple devices are made from common laboratory equipment to which the students have access. Students understand the concept and mathematical treatment of these parametres, now they have to build a device that will collect the required data. The class is encouraged to work as a group to brainstorm a design that will be used by each sub-group of two to four students. The device is simple, built around a filter funnel. After construction, students must determine various volumes before sediment samples are added. The sediment samples used for this are the bulk samples collected during week 2. The samples are added to the filter funnel, and compacted to approximately the field bulk density calculated in week 3.

Rigid wall permeametres are used to create sand packs for hydraulic testing using a constant head method. Groups of three or four students pack a permeametre with material collected during week An additional permeametre may be filled with another sand sample. For this additional permeametre, a commercial coarse blasting sand is used. This product is a relatively pure quartz sand article density = 2.65 g/cm^3). A sieve analysis is completed before the lab, and results are supplied to the students. The blasting sand is typically a very well sorted material, and should contrast with most natural materials. All permeametres are packed to achieve a bulk density equal to the average bulk density determined for the discrete samples collected from the cut face during week 2. If the bulk densities of the discrete samples vary significantly, the bulk density value for the bulk sample is used.

Now that the permeametres are constructed and hydraulic conductivity determined, they are used for studies of solute transport. Prior to the start of lab, the columns should be pre-saturated with a 0.003 M NaCl solution. It is also important to note here that it is best if solution is kept flowing through the columns as much as practical between lab sessions; otherwise, microbial growth may affect the initial K values that were determined in week 5.

Investigation of solute transport is easily accomplished by flowing the 0.003 M NaCl solution through the permeametre. The concentration of chloride in this solution is approximately 100 mg/L, which is easily detected using ion chromatography (IC) or ion selective electrode (ISE) methods. Higher concentrations may be used, but should not exceed approximately 500 mg/L to avoid significant changes in the density of the water. Transport phenomena are investigated by switching the source of water from the NaCl solution to a $NaNO_3$ or NaBr solution of similar concentration. The second solution is allowed to flow through the permeametre by switching inflow lines. Samples are collected for analysis by IC or ISE. The IC method is slower, taking about 6 to 8 minutes to analyse a single sample, but analyses give concentrations for both Cl and NO_3 (Br). Having information on bulk density and particle density for each permeametre, porosity may be calculated. Given the volume of the headers, the residence time for one pore volume is easily determined. In this lab session, students predict the time it will take for the second solution to break through, and may be guided to calculating the pore volume of the column. Through preliminary group discussions, students also evaluate other physical and chemical processes that may be occurring in their sand column. Factors such as dispersion and sorption need to be considered to make an accurate estimate of the breakthrough time. To limit the number of samples analysed and encourage students to think about the physical and chemical processes occurring in the column, students are given an analytical budget, thus limiting the number of analyses that may be performed. The group coming closest to catching the initial breakthrough is awarded a prize.

Hydraulic conductivity is determined in the field using percolation tests. These tests are commonly used in the industry for the permitting of on-site septic systems. Several accepted methods exist, and the specific method chosen may be that approved by individual state regulatory agencies. In this lab, we used a simple percolation test, which involves digging a shallow round hole approximately 10 cm in diametre. The sediment is initially saturated with water by repeated filling of the hole. Once an apparent constant percolation rate is observed, the hole is filled with water a final time. The hydraulic conductivity is determined by measuring the rate of water level drop in the hole. Given the diametre of the hole, the percolation rate is determined. This value may be compared with permeametre test completed in the lab on the same material during weeks 5 and 6.

The final lab session of this sequence provides an opportunity for students to catch up and compare results with their classmates. The focus for this session is to review the entire lab project and the data collected. Discussion may wander, but is guided toward assessing the potential aquifer in terms of its physical and hydraulic properties. Additionally, students determine which data are most relevant to their analyses.

The effectiveness of the laboratory project is assessed using four tools, pre- and post-testing, concept mapping, relevant questioning on exams and student feedback. Overall, results of this preliminary assessment indicate a positive impact on student learning. Due to the small sample size, no definitive analysis is possible until more data are collected from future classes.

Formal student assessment of the course using the IDEA tool, and anecdotal evidence suggest that students enjoy the hands-on laboratory project. Although some students had misgivings about the group aspect of the project, combining individual data into a single large data set, none suggested that this impacted their final grade. Overall, students expressed a feeling that they learned more in this type of laboratory setting than in more traditional labs, although there is an initial period of adjustment.

Despite the small sample size that prevents definitive analyses, the integrated field-laboratory project appears to be successful in addressing stated learning outcomes for the course, and in improving student understanding as demonstrated by performance on examinations and their ability to identify relationships between discrete hydrogeological concepts. Further benefit to the student is derived from the hands-on experience in both laboratory and field components as these experiences build a practical skill base. Finally, most students suggest that they prefer the overall format of the course laboratory. Their increased preference may be related to their increased understanding of the course material.

Urban heat islands are characterised by "islands" of warm surface air centred on urbanised landscapes and surrounded by progressively cooler air over suburban/rural areas. The urban-rural surface air temperature contrast varies diurnally and is one measure of the urban heat island (UHI) effect. Urban heat islands result from factors that differentiate the urban from the non-urbanised landscape, including: anthropogenic energy releases from heating-ventilation-air conditioning systems, energy emissions from industrial processes and motorised vehicles, the amount of available surface moisture and the differential heat capacities of urban building materials versus natural structures. The magnitude and spatial scale of UHI effects have increased in urban areas around the world and have implications for the transport of air pollutants, modification of bioclimatic conditions, heating and cooling costs and the confounding of long-term temperature records. The investigation of UHI effects provides numerous avenues of exploration that are both topical and relevant to students lives - characteristics that promote successful classroom implementation of science inquiry projects.

This chapter documents the process used for our Urban Heat Island classroom research project in our undergraduate "Atmospheric Interactions" course and discusses some of the outcomes both in terms of scientific results and achievements in student learning. The course is an upper division undergraduate general education science course that is taken primarily by students not majoring in science to fulfill degree requirements. The objective of the course is to study the physical and chemical properties of the atmosphere through lecture and field studies. We pay particular attention to the impact of human activities on the atmosphere.

This Urban Heat Island project is designed to parallel a scientific research process as it takes the students through literature review, preliminary studies, development of research questions, experimental design and implementation, data analysis and report writing. As an instructional approach it has the following features: it is situated in a meaningful real-world context; it provides students with an opportunity to deepen their understanding of physical principles through application; it requires classroom collaboration combined with significant individual effort; it incorporates an appropriate use of technology and it requires the communication of the analysis in the rhetorical style of a scientific paper. The design and implementation of this project is consistent with science inquiry pedagogy recommended by national organisations to be a part of undergraduate science education.

The Urban Heat Island Effect

There are numerous on-line and text references that provide good coverage of the basics of boundary layer meteorology and the urban heat island effect. The discussion below is based on lecture notes and the references noted above. It is provided as an introduction to the subject in order to understand the general design of the project.

The temperature of air near the surface, or boundary layer, is strongly influenced by the energy flux and physical characteristics of the surface. The natural energy balance between solar radiation input, long-wave emissivity and sensible heat transfer results in a diurnal cycle of heating and cooling of the earth's surface and atmospheric boundary layer. In contrast, significant diurnal variation in air temperature is not observed above the surface-influenced layer of the atmosphere or the free troposhere and above.

In order to examine the urban hear island effect, we can consider the energy fluxes through a shallow layer at the surface containing air and surface elements. The external energy source for this layer is solar radiation. The net radiative energy, Q_{net}, is controlled by the relative magnitude of shortwave and longwave radiation entering and leaving the layer. The radiative energy budget can be accounted for with Equation 1:

Although we pre-determine the overall research area (UHI effect) we attempt as much as possible to leave the design of the experiment and the specific research questions up to the students. Our role is to provide the students with enough background material and experiences that wild enable them to generate a well-designed project.

Each class session is roughly 2 hours in duration. Several of the activities are homework assignments that students are expected to complete outside of class.

The project is organised into five phases:

1. *Introduction of science content background relevant to the study (1 session).* Students are given lecture-based background information about forms of energy, energy transformations in the environment, temperature and fundamental physical properties of the environment. Supplemental reading in an atmospheric science text is also assigned. We use Earth Under Siege by Richard Turco since the rest of the course deals with air quality but treatment of these topics in a general meteorology text would suffice.
2. *Reading and analysis of a study published in the scientific literature (1 session)* A published research paper about urban heat islands is read and discussed in detail. We have used several different ones. Students are assigned the reading for homework and work in groups to prepare oral summaries of assigned sections. Through an in-depth analysis and discussion of these papers, we have an opportunity to examine how the physical principles of energy and energy transformations in various materials under many environments are key to interpreting data relevant to UHI effects. This literature analysis also introduces students to the rhetorical style and format of a scientific paper.
3. *Preliminary temperature studies (1 session) "T vs. The X Factor":* Preliminary studies are an important component of scientific research. (In fact, most experimental work can be categorised as such until we figure out how to carry out the most fruitful experiment.) During this phase of the project, students are asked to carry out simple temperature studies in the outdoor environment that would help the class to elucidate the effects of different environmental features (the "X - Factor") on measured air temperature. All hand-held temperature

measurements are made with an electronic temperature device (Check 1, Hanna Instruments - $25) that has been calibrated in water baths with National Institute of Standards and Technology (NIST) - traceable mercury thermometres. Each student checks out a thermometre for the duration of the project.

The goal is to carry out a study in which temperature varies with respect to one known variable. The assignment is titled "T vs. The X Factor". Examples of such studies are: measuring temperature as a function of: distance away from a river, elevation (by climbing a hill or a parking structure), distance away from a building or ground surfaces of varying kinds.

We usually ask for one volunteer to study the effect of air speed on temperatures by measuring temperature out of a car window at various speeds. A second "volunteer" would be asked to conduct a similar study but with a thin layer of wet tissue around the probe. Without exception, students have come into the class believing that increased air speed alone will reduce the temperature sensed by the instrument and are surprised a second time by the effect of the wet tissue.

This assignment is designed to allow the students to gain experience in scientific claimsmaking. Students present their experiments to the class and propose explanations for the data. Obviously, in "X factor" studies, students are never completely able to isolate the temperature variation to one variable. Unlike canonical, well-controlled teaching laboratories, these temperature measurements are a function of numerous variables. In their presentations students are forced to account for and/or rationalise the effects of "unintended" X factors on their temperature measurements.

4. *Preliminary research questions, design of measurement protocols and transects (1 session):* Based on their reading and research, we attempt to develop researchable questions constrained by the limitations of resources temperature probes) and time limitations that are well-known to experimentalists. Examples of such questions are: "Are there spatial differences in air temperature that are related to level of urbanisation? ", "How does the presence of rivers effect urban climatology?", "What is the influence of vegetative cover on the ambient

temperature?" During this session we also discuss some of the strengths and weaknesses of previous class studies in order to improve on experimental design. For instance, during our earliest studies we found that data collection was too sparse along each transect. Subsequent groups have designed their studies to collect 15-20 data points along a transect traversed within an hour.

Once preliminary questions are agreed upon, students determine what data is necessary to address the question. This process leads to the determination of measurement protocols and ancillary data collection. This ancillary data collection protocol probably requires the most professional intervention, as inexperienced researchers it is difficult for students to anticipate what data might be useful in subsequent analysis (for instance cloud cover, surface characteristics, wind speed etc.). However, once these factors are considered, it is useful for students to struggle with the "calibration" of ancillary data. For instance, how will each student determine level of cloud cover? Decisions are also made about when to make the measurements, which days) and at what time(s). Discussion can often become quite animated as students come to a consensus about how to conduct the study. Measurement protocols are based on their previous work and experience with the preliminary studies. The students grapple with the question, "How can we make meaningful temperature measurements in a complex environment in order to investigate the research question at hand?"

Typical protocols for temperature measurement:

1. Hold calibrated probe one approximately lm above ground and arm's length away from body.
2. Do not make measurements within 3 feet of a building.
3. Shade probe with an umbrella or other suitable device.
4. Wait at least 30 seconds and for thermometre display to stabilise.
5. Record temperature reading and other data.

Most groups determine that the optimal time of day for a maximum UHI effect is right before sunrise. At the end of the session, each student leaves with data sheets and knows when the data collection is to be made. In addition to the temperature measurements made along transects, there are two important measurements necessary for making sense of the field data in later analysis: 1) an assessment of the stability of the air and 2) a continuous temperature record at a fixed location. Both of these measurement tasks are the responsibilities of a student or a student group.

The UHI effect is maximised under conditions of low wind speed and high atmospheric stability (temperature inversion) since this minimises mixing through vertical or horizontal motion. It is useful to document these conditions during the period of data collection to help explain and interpret the data. Although the National Weather Service provides twice-daily vertical profiles of temperature and wind speed, the National Weather service stations tend to be away from urban areas. Another source for data is a relatively new product derived from GOES satellite soundings but cloud cover limits availability of useful data.

Stability conditions can be also be successfully assessed using, for example, a small rocket launched with temperature and pressure logger payload (we used HOBO, Inc. sensors and dataloggers), generating smoke and taking pictures of its movements, or taking temperature measurements along elevation features (such as hills, towers or tall buildings). Relative stability can be determined by comparing temperature profiles with adiabatic lapse rates.

Since the field measurements will be made along a transect during a window of time, it is necessary to have a continuous record of temperature made at one fixed station, preferably in either the coolest or warmest region of the study domain. With this continuous record the temperature value collected in the field, which should be timestamped, can be subtracted from the fixed station temperature at that time. The temperature difference or UHI effect can be calculated for each data point at any time. For example, if a measurement was made in the field at 6:05 AM, the UHI effect would be calculated by subtracting the temperature at the fixed site at 6:05 AM from the field measurement. Continuous temperature records can be obtained with temperature sensors and data loggers (such the ones used for rocket launches) or from airport records. This approach will generate a UHI "snapshot" with an "exposure time" of the monitoring window.

5. *Data analysis and report writing (3 sessions):* Students generate a common spreadsheet that contains the data from all participants. This spreadsheet is used to calculate the UHI by subtracting each of the temperature values from a downtown meteorological station.

The UHI value is calculated by subtracting the temperature measurement made along a transect from the fixed station measurement at that time. In the experiment, "Benson" is the temperature at the fixed "Benson" site, where a temperature logger was located. Note that ancillary data, such as observed land use and cloud cover, that the students had agreed to collect are also part of the data record. The X and Y coordinates were determined using mapping software but could also be determined using mobile GPS units carried by students. Once the collected data are merged into a common dataset, students are charged individually to analyse the data and write up their analysis. This has been our practice but small group analysis of the dataset is also possible.

Data analysis can be accomplished with or without sophisticated software tools. If spatial analysis software is unavailable, data can be analysed by students mapping the data by hand onto thematic maps of urban regions. Thematic maps based on land use, vegetative cover, topography etc. are generally available at city planning agencies or local USGS offices.

We have found that Geographical Information Systems (GIS) software is well-suited for the analysis of the data generated in this project. One popular publisher of GIS software, ESRI Inc., provides a freely available and powerful GIS viewer and analytical tool on their website called ArcExplorer. Use of GIS software for spatial analysis allows students to assess, quickly and visually, the relationship of the data to various spatially varying parametres. This software program has additional analytical tools, such as the ability to query the data with Boolean-like operators, which allows the analyst to screen the data for various conditions. For instance, one could examine only data for which cloud cover was less than 10% and measure their distance from the city centre.

There are two barriers to simple implementation of the ArcExplorer for this project. First, this package does not allow the creation of new spatial datasets. In order to convert the data from spreadsheet format to a format that is GIS compatible, it is necessary to utilise the commercial package ArcView. ArcView is widely available on university campuses (and some middle and high schools) and the conversion process only takes a few minutes. A friendly colleague or graduate student in the Geography department is likely to do it for a cup of coffee. The second barrier is the availability of suitable spatial databases to overlay with temperature data. Increasingly, the federal government, municipal agencies, and others are providing their spatial data in GIS compatible form. Students have three in-class work sessions in which to analyse the data with respect to their research question. We have found that allowing work time during class time" is essential as it provides an opportunity for collaboration among students, technology access and assistance, and quick feedback from instructors. Each student is responsible for writing his/her own paper based on the collective data. The paper is written in the form of a scientific paper and the students use our reference UHI paper, analysed in step 2, as a guide.

We have conducted this chapter with several classes over the last three years. This particular data set of over 200 points was collected with a class of 22 students. The data were gathered between 6:15-8:15 AM on November 10, 1999 using protocols discussed above. Each student was responsible for gathering data along an assigned transect, which was determined by the class. This particular dataset was acquired in the second attempt. For the first scheduled data campaign, cloud cover appeared overnight and did not generate a "good" dataset for analysis. The students, with a little prompting, decided to repeat the experiment under better conditions.

The overarching analytical objective for the students is to explain the pattern of temperature variation present on the map. Students are provided with access to several spatial databases for the region that could help their interpretation. These spatial databases can be imported into ArcExplorer include: land use, zoning, elevation, vegetative cover, soil type, population, etc. Although these databases have been organised by our regional government for research purposes, many of these databases are available on-line at federal and local government GIS sites.

Once these maps are generated, students must devise methods of quantifying the data for comparison. Students who first generated this graph were disappointed that all the data "did not lie on a straight line. The challenge in working with this kind of data with students is to get them to see through the "messiness" of the data with statistical or graphical tools and to persevere in examining the reasons for the underlying variations. Although UHI is plotted against distance many other variables are at play and students struggle with the multi-variable nature of environmental research. Students submit a draft version of their report for review. The draft-revision process is critical in engaging the students in the desired learning outcomes. Most student papers improve dramatically in this process.

- understands the factors controlling the energy balance at the earth's surface
- understands how the vertical temperature profile of the atmosphere relates to atmospheric stability
- can interpret indicators of atmospheric stability
- understands how atmospheric stability effects UHI
- can explain the diurnal temperature cycle
- can explain how human-made structures and water effect the diurnal cycle

Science Process:

- Recognises the importance of controlling variables and accounting for variables that cannot be controlled
- able to frame addressable research questions
- understands the role of preliminary studies in research
- understands the relevance of precision and accuracy able to construct a supportable knowledge claim

Skills:

- able to gather and record data using a scientific instrument using data collection protocols
- able to graph data using Excel
- able to utilise ArcExplorer to map data and thematic overlays

In addition to the learning outcomes outlined above, we have also observed two important attitudinal shifts in our classes: 1) Recognising the importance of collaboration in science and 2) Becoming aware that they can "do" science. The design of the project makes every student's data collection activity critical to the research. In fact, the spatial analysis is nearly impossible without high-resolution data. Students go to great lengths to insure that their transect is completed at the correct time. During the data analysis and report writing phase of the projects, many students experience frustration with inconclusive data" as they search for the "right" answer. As students persevere through this period they come to Recognise that doing science is usually not about finding a clear answer but about discerning meaningful and supportable patterns in otherwise "noisy" information. Follow-up projects have included the economics of heating and cooling with an UHI-affected area, assessing the UHI with remote temperature data and analysing the conditions for optimal UHI effects.

Science education reform efforts underscore the need to engage students at all levels in "doing science as science is done". The science inquiry project described here presents a transferable opportunity to engage students in an authentic investigation that incorporates real world scientific processes. The project goals are consistent with the goals of science education reform efforts for undergraduate science courses. Student enrolled in a general education science course undertake the evaluation of the urban heat island effect in the Portland metropolitan area. We have implemented this project with classes of between 20-35 students and with over 150 students altogether. This project successfully utilises the collective abilities of the students to design and implement a scientific experiment and engages the individual student's ability to analyse the data and communicate the results.

Geologic Problem Solving

Field-based instruction is widely acknowledged to be a central part of undergraduate education in the geological sciences. Representatives from the energy and environmental industries as well as academia and government gathered at the Indiana University Geological Field Station (IUGFS) in a community wide meeting on the future of field-based education in August of 2006, universally (but anecdotally) acknowledged the central importance of undergraduate field education, especially in capstone field camp settings. Even when graduates of these programs went into lines of work that did not require field skills per se, those present at the meeting noted that problem solving skills and habits of mind such as actively using multiple working hypotheses were greatly increased from geologic field camp education.

The geological sciences are among the most visually oriented of all sciences, and the earth sciences as a collection of allied fields collectively relies on spatially integrated and spatially embedded data more than any other branch of the natural sciences. The field clearly makes these spatial relations more concrete, but also demands abstract spatial thinking as well. Only recently have science education researchers begun to focus on the unique aspects of problem solving and cognition, and geologically specific spatial abilities that characterise expertise in this area, particularly in the critical outdoor teaching, learning and research environment.

The outdoor field-based environment is not ideally suited for controlled tests of cognition as would normally be carried out in laboratory or classroom educational setting, as the variables involved are many and human factors (i.e. human response to terrain, exhaustion, discomfort, etc.) become involved. Therefore the study of problem solving skills needs to be treated by proxy measures, and needs to explicitly work with the study of problem solving and decision making as it happens in natural, real-world settings. We have developed a methodology for analysing navigational choices recorded on GFS units worn by students during field examinations that we have demonstrated reflects problem solving stages as defined by some workers in the cognitive science research fields of Naturalistic Decision Making. We discuss this approach and our methodologies in the sections that follow.

Field-Based Learning in Deformed Sedimentary Sequences

Most field camp courses have common curricular components that have students working with deformed sedimentary units. While field studies in metamorphic or igneous terranee are equally valuable to the well-rounded education of any undergraduate geologist, optimal controls for educational research of the type presented here are difficult to establish. Fortunately, a large portion of field education is based on structural and sedimentological problems in sedimentary rocks, and many of the field areas used for the instruction of undergraduate geology majors share a number of features which provide constraints which facilitate the study of student learning. These kinds of problems tend to have a highly deterministic geometry which allows prediction of subsurface structure from surface information, and prediction of the likely surface exposure in asof-yet unmapped areas. This type of geologic problem lends itself well to testing by multiple geometric working hypotheses that can be tested by planned traverses of a field area optimised to search for data that confirms or rejects hypotheses. The proposition that underlies our research approach is that the navigation decisions made by students while investigating this type of field problem reflect their internal problem solving approaches as they fit testing and verification strategies derived from their mental models to traverse plans.

Problem Solving

A central issue emerges in the framing of this research, namely what is problem solving? And what is geologic problem solving relative to other kinds of problem solving in science? In many sciences and in mathematics, especially at the undergraduate level, students are confronted with analytical problems, where they are basically given puzzles to solve. In front of them lies all of the necessary data to completely solve the problem, and the primary challenge to the learner is to draw on their recognition of common patterns to put these pieces together to find the solution. Pattern recognition is a hallmark of "expert behaviour" as described in Bransford, Brown, & Cocking (2000), but the primary distinction here is that all data is available at the outset.

Geologic problem solving occasionally does involve this style of investigation, but in the field this is rare. Field problem solving involves a full range of navigational skills and map reading and many discrete skills related to spatial visualisation. Beyond having to bring many separate skills to bear on geologic mapping situations, and assuming even complete mastery of these requisite skills, problem solving in the field also involves operating in fundamentally data-poor and underdetermined situations. Researchers (and students who function as researchers during exercises and field exams) never have all the information they need to fully solve any given problem with the confidence of an analytical solution. They must rely on the construction of multiple working hypotheses which can be pursued in order to gather more relevant data which in turn improves the working hypotheses. This iterative, boot-strapping situation does also depend on expertise, but involves additional skills of planning and field navigation designed to optimise a path through likely data-rich regions in a field area given ever-present time constraints. Because of these additional features to problem solving in a geologic field context, it is clear that the act of problem solving is best studied in its naturalistic context and is impossible to study and duplicate in a fully controlled laboratory because so much of the problem solving strategy is bound up in individual response to the real situation. This is what leads us to the research traditions of Naturalistic Decision Making, and the conclusions that lead from this approach that allow us to understand how individuals of varying expertise tend to behave and are best taught to respond to variable field conditions.

Naturalistic Decision Making

Our theoretical approach to this chapter is rooted in the body of research in the cognitive science field of Naturalistic Decision Making (NDM). This area of research deals with problem solving and decision making in data-poor situations, usually under time constraints, where the presence of expertise has a strong influence on moment-to-moment decisions made by problem solvers. NDM researchers typically study the real-time decision making pathways and processes of professionals such as firefighters, military commanders, air traffic controllers, manufacturing production managers and any other class of decision makers who have to make relatively rapid decisions with incomplete information. In data-poor situations, success relies on matching multiple emergent patterns to data as it comes in. Expert problem solvers employ pattern recognition to make an educated guess at the "class" or "style" of a situation and make decisions for gathering additional data that quickly reduce the number of possible solutions and constrain the true nature of the problem. In a geologic setting, expertise is required to visualise and interpret these features from exposed features, and the evolution of novices into experts in this field is poorly understood. Studies of teaching and learning in the field, must also consider the complex interactions of factors that may have a bearing on an individual student's actions, decisions, mental model formation, and ultimate learning outcomes.

NDM is an appropriate context for geoscience education research especially in a field setting because it places an emphasis not only on the cognitive tasks and actions of an individual or group, but also on the context in which that person or group of people is responding to the task at hand - in this case the geology, exposure and topography of the field area itself. Furthermore, this area of decision-making research is all the more appropriate to the study of field-based geologic problem solving because of its emphasis on systematic observations and analysis of skUled decision makers in naturalistic, i.e. not experimental, conditions. Experimental settings are often not able to adequately reproduce all aspects of a field situation that affect real-time cognition and decision making. This general area of research is more suited to decision making processes informed by intuition rooted in expertise rather than analytical, algorithmic, or optimisation types of solutions. This captures much of the difference between geological investigations, exercises and examinations in the field at the advanced undergraduate level as compared with physics or chemistry problem solving at this level.

The recent work published by Claire Bond and colleagues in GSA Today underscores the central role of expertise in interpretation of geologic data. Their work shows that experience can strongly bias interpretations of incomplete data, in then case a synthetic seismic data profile generated from a 2D forward model of deformation of sedimentary strata. The researchers knew the "right" answer - a situation almost never encountered under real field or exploration conditions - and had over 400 volunteer geoscientists with some level of experience in seismic interpretation produce their best interpretation of the work. Not surprisingly, experts carried their past experience into their problem-solving processes, but interestingly, those experts who drew on many experiences and techniques useful in multiple contexts tended to produce better answers. While this chapter does not bear directly on our results, it is absolute confirmation that the assumptions and approaches of Naturalistic Decision Making are valid and useful in understanding the work of professional geoscientists in field or exploration settings, and in developing best practices in undergraduate education that lead to that expertise.

Many process models have been proposed within the NDM tradition. Out of these characterisations of problem solving and decision making in naturalistic settings, we find that the schema model of Marshall (1995) to be most productively adapted to geologic problem solving. This model Recognises that decision making involves the construction of mental models (schema) that are in turn constructed of subordinate schema that work together iteratively to provide the basis for decision making.

Marshall's research group worked primarily with U.S. Navy battlefield tacticians whose interface with the units on the ground was a large computer display station and active command and control communications. Subjects were monitored by eye-tracking equipment, which recorded the path that tacticians' eyes followed across the displays during various scenarios. Coupled with this was analysis of communications, subsequent interviews, and in more recent work, pupil dilation vibration rate, which the group showed to be related to concentration intensity. From analysis of eye tracking patterns correlated with concentration intensity and other communications evidence of effective (or not effective) problem solving, Marshall's group produced a model with four components that iteratively work together to construct problem solving in these types of datapoor, time-limited situations that they called a schema model, based on the schema or mental submodels that had to go into the overall problem solving process.

Marshall's conceptualisation of schema in problem solving includes the following four elements:

Identification Knowledge: The ability to Recognise relevant information and assess from clues in the environment when a situation is similar to prior experiences or education

Elaboration Knowledge: The immediate associated recall of related facts and elements which aid in the confirmation or adjustment of the initial assessment of a situation from identification knowledge. This is similar to the "chunking" of information common to experts reviewed in Bransford et al. (2000).

Planning Knowledge: The ability to draw inferences and estimates, create goals and plans using the framework provided by identification and elaboration knowledge.

Execution knowledge: the ability to utilise skills and procedures as needed to provide further information or take additional action to further a solution.

These are iterative, interactive portions of the larger mental model (schema) that a decision maker uses to Recognise emerging situations and direct current actions and future data collection priorities, but the absence of any of them prevents effective problem solving. This is easily applied to geologic problem solving, especially in field mapping and structural problem solving under time constraints. In the field, geologists identify rocks and make relevant measurements, elaborate through multiple working hypotheses explaining how these data are fit by larger-scale solutions, make plans to traverse the landscape to most efficiently test these hypotheses, and then execute the plans safely as terrain conditions allow. Clearly these individual steps are repeated as needed at many temporal and spatial scales as new data is revealed during a field traverse.

From an educational research perspective, typically only the identification and execution steps are easily externalised. From checking completed geologic maps, notes and other direct observations in the field, it is clear if an identification is correct, and by tracking navigation, the execution step is recorded. Close analysis of patterns in the navigation data, along with field notes and finished geologic maps can be used to infer the quality of elaboration and planning knowledge in a field teaching setting, but we will show in this chapter that navigational patterns can also be analysed to shed light on these otherwise internal, mental processes.

This study was conducted in an Advanced Field Geology course for undergraduate geology majors, conducted in the contractional belt of the northwestern San Gabriel Mountains near Frazier Park, CA. Students in this course had completed beginning and intermediate semester-length field courses and were all long-time residents of Southern California and were accustomed to steep topography, typical field conditions in the region and the general geologic and tectonic history of the region. The first author was the also the instructor for this course, and had been the instructor for many of these students in their introductory field course, so care was taken to secure informed consent between the second author and student participants in compliance with our IRB approval. The first author had no knowledge of which students had elected to participate in the study in advance, and no data analysis was conducted by anyone on the research team until after the course was completed. We acknowledge that these authors did have insights into individual student histories and tendencies which likely influenced some of our interpretations, although as will be shown below, efforts were made to reduce this effect by triangulating objective and subjective measures in forming conclusions.

Students worked in pairs or groups of 3 for an exercise in a given region for a week, and then completed an all-day independent field examination in a nearby location, in the same sequence of rock units and exhibiting a similar structural style. Students were instructed not to communicate or follow one another, and were monitored for compliance. Students had approximately 7 hours to produce a geologic map from available field data in a bounded region roughly 2 km^2 in size. The area included sufficient exposure of deformed sedimentary rocks to find geological data but also had widespread tree cover and very steep topography at the southern boundary. These features necessitated intelligent planning of traverses across this landscape to maximise data collection and interpretation.

Students were fitted with GPS units set to record their location every 3 minutes for the entire test period. They were allowed to look at these units, but their base maps deliberately had no geo-referencing information, rendering the units useless for navigation. Students were also instructed to make stations at data collection locations and record raw data both on the base maps and in their field notes. By the end of the test period the students were expected to hand in a completed geologic map of the region along with notes containing their raw data and illustrating their ongoing thought processes throughout the exam. We were able to recover and analyse data from only eight of our 15 participants in this field area due to loss of satellite coverage or power failure for some GPS units.

We scored all of the student maps against a traditional rubric constructed for evaluating geologic maps in field instruction, referenced to a map constructed by course instructors for this examination area.

Points were awarded for accurate recognition and placement on the map of key geologic features such as structural elements (e.g. fold axes, etc.), contacts between geologic units, and correct identification of formations. Decreasing amounts of points for each key feature were granted with decreasing accuracy of location or omission of that feature.

All complete GPS records were imported into ArcGIS for analysis. Density clustering produced "hotspots", or contoured plots showing frequently visited locations where students had stopped throughout their traverses. The field area was later rechecked by investigators, and we discovered that these hotspot locations did indeed correlate with key geologic features. It is clear that these often-visited locations were necessary stations for students to visit in order to gather critical data. This does not mean that students were able to understand the significance of what they had found at these locations, but it does show that students were drawn to the same significant features. This observation is explored further below in the general discussion of tracks and track analysis.

We constructed polygons of each consecutive five GPS data points, representing 15 minutes of work for that student. Adjacent polygons were strung together in a sequence to create time-series tracks. This data-processing approach enabled two kinds of complementary analyses. Polygon scores were assigned based on the degree to which polygons intersected hotspots, and a second coding scale that captured the dynamic qualities of the tracks. The polygon scoring scale is a rubric-style measure that runs from zero to three depending mostly on the degree and duration of intersection of a student's navigation track with known critical locations (hotspots). A score of 3 means that the polygon completely engulfed a hotspot, indicating that the student stopped and investigated this region. Scores decrease with evidence that a student made a glancing pass of a critical area or traversed through a critical area too fast to be confident they meaningfully investigated the region. A score of 0 was assigned for the polygons that did not intersect hotspots at all. Students were only scored once for duplicate passes across locations.

To capture and describe the temporal dynamics of the track sequences, we settled on a two-part coding scheme that lent direct insight into the dimensions of Marshall's schema model that undergird our study. The four primary codes are related to travel speed, and each polygon in the time sequence was coded as fast, measured, slow or static. The code choice was tempered by knowledge of the terrain, such that a small polygon on steep terrain could be lebelled as fast if it was relatively larger than other nearby polygons. The six secondary codes were related to sequences of polygons and reoccupation of sites in the field area. This approach enabled us to code track sequences such as "double back" maneuvers or star-shaped sets of polygons with a common origin that suggested repeated investigatory forays from a single starting point in a region, called a "touch and go". We coded these in time-series fashion from start to finish of the field examination, and totalled the number of instances of each code for each student.

Semi-structured, stimulated-recall style interviews were conducted with selected students, and they were asked to explain their mapping sequence, results, and thought processes using their finished map and notes. These interviews and analyses are reported in Lieder (2005) and Lieder and Riggs (2004). The students were shown their traverse patterns and the hotspots. Students were asked to talk about what they investigated throughout the exam, and were probed to elucidate their problem solving techniques. Interviews allowed us to independently calibrate our interpretations of track patterns and our track coding to the students' account of their problem solving, and allowed us to understand which aspects of the coding scheme indicated success or failure of the various stages of problem solving.

We present below all the individual results from this investigation, including each complete polygon track, the corresponding coding timeseries (each polygon is shown by the range of the 5 GPS points used to assemble it, 1-5, 6-10, etc.), and their finished geologic maps. Unfortunately, the print medium does not permit the dynamic presentation of all students' polygon tracks accompanied by real-time coding illustration. To augment this static presentation of our data, we have posted animated versions of these figures on the Internet which show the temporal progression of each student's traverse. Below we have attempted to provide a relevant summary narrative for each student in the study, with analysis of relevant major points observed in each students' behaviour in the field.

Analysis of Adrianne's polygon track produced very few secondary codes and the primary codes indicated that Adrianne alternated between fast and slow paces. The "Path Cross" code appeared four times and each time was paired with a "fast" pace primary code indicating that she moves through these reoccupied areas at a fast pace, only doing so to cross into new unoccupied areas.

Adrianne's overall traverse in the field included several looping maneuvers and this strategy led to the occurrence of these path crosses. It's also clear from the tracks that Adrianne spends a significant amount of time outside of the field exam area and this doesn't seem to affect her performance as indicated by her map score as she received the highest map score. As shown by the lack of secondary codes and her completed map, Adrianne maximises her time in the field area and a majority of her time was spent near significant geologic structures which she correctly identifies.

Jay had one of the most orderly traverses through the field as he completed one full clockwise circuit. Jay's pace was either measured or slow, until the end of his traverse when it became fast. There were very few secondary codes throughout most of the track until the very end, which coincides with the increased pace as noted. The secondary codes include "back and forth", "retrace", and a "double back", which all represent maneuvers in which he reoccupies areas that he had been in at the beginning of his traverse. All but one of these codes occurs after Jay has already made a full traverse of the entire area; the one code that occurs before this is a double back. Also at the beginning of Jay's traverse he uses approximately 45 minutes to presumably get to high topography. Jay was accountable for all the major structures in the field, which is likely related to the efficient path that Jay takes through the field, but he was marked off on his completed map because the placement of the structures were inaccurate.

Similar to Jay, Julie had a very orderly traverse that was one large clockwise loop through the field area. Julie's pace was very measured or slow and there were several occasions where she was static for ~10-25 minutes at a time. Like Jay, Julie had very few secondary codes indicating a reoccupation of an area and these didn't appear till the very end of the traverse; one of the retrace and double back codes are proceeded by out of bounds maneuvers. Julie was also able to identify all the major structures in the area, but she struggled with the placement, trend, and sense of these structures accounting for the point deductions on her completed map.

Bob initially starts out in a clockwise circular pattern, but his traverse quickly becomes inefficient as he executes a zigzag pattern up and down steep terrain throughout the field area. Overall Bob's pace was evenly distributed between all 4 primary codes, although in the beginning his pace was mostly slow to moderate, while at the end of his traverse it was fast or static. Bob had several secondary codes which start approximately half way through his traverse and after he has started the zigzag portion of his traverse. The touch and go codes and double back codes are a result of Bob's zigzag pattern as he returns to the valley before heading back up the mountain and back down. The retrace codes associated with Bob's traverse are at the very end of his traverse as he returns to the beginning where he remains static, before returning to the initial starting point. Like his higher scoring peers, Bob identifies all the important structures, but his lack of precession in placing the structures and identifying the correct sense of displacement leads to a reduction in points.

Bill's traverse is the only non-circular/loop pattern that appeared to be planned as such. Bill's traverse looks like a 3 pronged approach as he starts out by heading west, heads uphill, comes back down to the valley and heads to the far east, where he heads up terrain again before back coming down, heads down the valley, stops halfway between the previous 2 approaches and heads uphill, before finally returning to his starting point. Bill is constantly moving as he is only static at the beginning and the very end and his pace is evenly distributed between slow, moderate, and fast. The majority of Bill's secondary codes (Retraces and several Double backs) are the result of Bill's planned navigation pattern. Bill is the first student to not locate all the important geologic features and this is reflected in his map score; his lower score is also the result of his failure to accurately locate the features he did identify on the map. His failure to locate all the structures is likely a result of his navigation pattern, because he did not visit areas that the others did who had more complete patterns.

Jesus had one of the most complex navigation patterns (right) and his maneuvers appear relatively random and unplanned. Jesus' pace was generally very slow in the beginning and end, while his pace was rapid in the middle of his traverse. Jesus had the highest amount of secondary codes (tied with Mark) spread out evenly throughout the navigation; the high number of secondary codes are likely related to the complex, random, and unplanned nature of Jesus' traverse. Jesus also had the highest variety of codes (double back, retrace, branch, touch and go, and path cross), which is also related to the complexity of his navigation pattern. Significantly, he also has a one-hour static period near the end of the examination time spent near the isolated data point on his map in the northeast sector of the field area, suggesting a total state of confusion and/or a subject that has completely given up on the exam. This is also seen in the work of Marshall (2002, and personal communication) in eye tracking paths of battlefield tacticians on tactical displays who have effectively surrendered or become mired in confusion. Jesus' below average score is a result of his failure to identify the fold in the valley or the tear faults in the southern portion of the field exam area. Looking at Jesus' tracks it's evident that he did not visit the areas where these structures were.

Mark's traverse looks similar to circular/ loop patterns of the higher scoring students, but is smaller and is not efficient. Mark's primary codes indicate that he moved at a rapid pace (the fastest of anyone). Most of his secondary codes occurred towards the end of his traverse and look to be related problem solving issues as he continually reoccupies the valley for the last 2+ hrs of his traverse. This rapid back-and-forth movement over previously investigated ground has also been seen commonly in Marshall's eye tracking work (2002 and personal communication), and is interpreted as a sign of confusion. Subjects in her studies will commonly glance back and forth repeatedly over known data they do not understand, hoping to literally see something new. Mark has demonstrated precisely the same behaviour, but by walking repeatedly over the same ground on a much larger physical scale than any eye tracking exercise. Mark has a couple of secondary codes at the beginning of his traverse, but these are more likely related to terrain issues than problem solving ones. Mark's low score on the exam is because he didn't identify the tear faults and the location of the thrust fault is poorly placed, this is likely related to his lack of time spent where those structures are located.

Jesse had a circular/ loop pattern through part of the field area. Jesse's primary codes indicate that he moved very slowly through the field area and in fact had the slowest pace. Most of Jesse's secondary codes occur at the beginning and the end when he's not moving in a looping traverse. Even though his traverse is similar to some of the higher scoring students, it is less efficient as he doubles back and retraces at the beginning and ends causing the loop to "pinch off", which causes Jesse to miss large sections of the field exam area. Jesse missed several key structures, including the east tear fault, the fold, and the majority of the thrust, leading to a severe reduction in his map score.

We found only a loose correlation of map and polygon scores in our data, and there were many instances where students with the same polygon score had widely varying map scores, suggesting that this measure alone did not explain the range of performance observed. However, the two-part track coding scheme presented allowed a finergrained analysis of the navigation tracks and yielded many insights into problem solving steps and helped more completely link navigation characteristics to performance.

As can be seen, a lack of secondary codes is a substantial predictor of high performance, especially when the primary code is taken into consideration. For example, the fact that Adrianne's path cross codes were paired with a fast-paced primary code, shows that even though she did spend time in previously visited areas she spent very little time there. Adrianne's path crosses indicate that she may not have initially had a navigational plan, but the low amount of time spent reoccupying areas indicate that she was successfully able to identify and elaborate on information in those initial areas and permitted her to quickly form a plan, thus carrying out the planning and execution steps. The successful implementation of the full schema framework is reflected in Adrianne's map score. In contrast, the nature of Jesus' path crosses indicates a deeper problem solving issue. The difference is inferred from the student's overall pattern: Adrianne's organised traverse versus' Jesus' random traverse.

Further insight to the application of Marshall's schema model is given by the generally high polygon scores and the coincidence of hotspots with significant geologic features. Students were monitored by the authors during test conditions to make sure they were not following one another or collaborating, and the terrain and vegetation cover in this area is sufficiently complex and dense to prevent casual observation of one student's movements by another. We are reasonably confident that students reached these locations on their own guidance and stopped at these locations because of their own recognition of geologically significant information at these locations. This represents successful completion of the identification step in naturalistic decision making by virtually all of the students, which would be expected of advanced geology undergraduates with the amount of prior field experience these students had. The large variance in map scores, coupled with insights from our track sequence coding, indicates that the subsequent failure or success of problem solving follows largely from a failure to elaborate successfully on the information gained, and/ or a misinterpretation of the information gained at these sites that led to erroneous elaboration into mental models of the field area that could not be substantiated with further investigation. Student interviews and analysis of notes supports this interpretation, as those students who had poor final maps also tended to have little evidence of planning and model elaboration in their recorded field notes.

A distinction should be made between those students without clear, emergent navigation plans, the ones with efficient plans, and those with inefficient plans. The appearance of a plan does not necessarily correlate to above average scores; Jay, Julie and possibly Adrianne, who all evidently developed navigation plans scored above average, but in contrast students like Bill, Bob, Jesse, and Mark who also had plans, performed below average because their plans were also largely inefficient. Certain types of traverses will cover just as much area as others, but will be physically exhausting which could possibly tax mental facilities, leading to a decreased map score. The formulation of an inefficient plan or no plan at all (Jesus) represents a breakdown in the stages of problem solving. Analysis of the individual secondary codes reveals that significant amounts of time (in some cases hours) were spent by these 5 students reoccupying areas. Additionally, the occurrence and repetition of certain codes (back and forth and retrace) indicates a decrease in problem solving. For instance, Mark executes several back and forth maneuvers in the same area at the end of his traverse, suggesting that he was struggling with the geology in that area. There is a loose correlation between the number of secondary codes a student has and performance. The top 4 map scorers had the lowest 4 secondary codes except for one instance (Jay), and the bottom 4 map scorers had the highest 4 secondary codes except for one instance (Bill).

General Conclusions from Track Analysis:

- Secondary codes are a result of a student's plan (or lack thereof) in the field, which could indicate how much active problem solving they are engaged in. In some cases, like Adrianne, path crosses do not seem to indicate a lack of problem solving ability, but just indicate a lack of an initial or consistent plan. In contrast, the nature of Jesus' path crosses indicates deeper difficulties with problem solving. The difference is inferred from the student's overall pattern: Adrianne's organised traverse versus Jesus' very random traverse.
- Certain types of traverses will cover just as much area as others, but will be physically exhausting which could possibly tax mental facilities such as problem solving leading to a decreased map score. For example, Bob visits the same areas as his higher scoring peers, but he receives a lower score on his map.
- The occurrence of certain codes (back and forth), or the repetition of a code, may indicate a decrease in problem solving ability. For instance, Mark executes several back and forth maneuvers in the same area at the end of his traverse,

suggesting that he was struggling with the geology in that area; this is supported by the lack of structures and inaccuracy of his map in this area. According to Marshall (2002 and personal communication), this type of rapid back-and forth movement in her eye tracking studies is also indicative of confusion, which establishes an interesting parallel between her studies of behaviour and cognition and this work. Discussions with the Marshall research group about this work have also revealed striking similarities with patterns developed in eye tracking research that we have not discussed here. Our methods and approaches with track coding and polygon scoring have already flowed back from this chapter in to that research area, and the unexpected similarity of eye tracking and field navigation patterns with similar problem solving task characteristics but vastly different temporal and spatial scales offers a new comparative window into basic problem solving cognition and the role of expertise.

Limitations and Questions

This chapter contributes new, independent tools for measuring and monitoring field geology performance that do correlate with traditional map grading and which give higher resolution understanding of specific difficulties to geoscience education researchers working with field based learning at this advanced undergraduate level.

However, there are a number of questions raised by this data set that we are continuing to investigate and which call for further investigations of this nature to be designed and conducted. A range of methodological, technical and research questions for future investigation are raised by this chapter, including:

- How are navigation patterns effected by classroom educational background (prior preparation), prior field experience, personal background (outdoor comfort/experience), and social influences (used to group/ independent work)?
- What does expert behaviour look like using this same methodology? We are actively analysing new data with beginning students, but it would be useful to know what the expert "goal state" looks like to calibrate our measurements.
- Are issues of Novelty Space, the "newness" of a field camp setting relevant to performance in field examinations?

- Human factors (athletic performance and related physiological issues) are likely relevant to this analysis, but we have no measure of these factors at this time other than our personal knowledge of the students involved in this chapter.
- While we now have a good framework for forming research questions and analytical inquiries using this of this kind of data, we do not yet have a rich enough data set to test our methods for sensitivity in sampling rate, polygon methodology, and trustworthiness in primary and secondary maneuver codes. Deeper qualitative interviews and observations are needed, as are more advanced quantitative methods and larger data sets with a broader range of student participants.
- We have no way to independently or quantitatively gauge how topography in the test area influenced navigation or traverse speed - additional well-known map test areas are needed to help constrain topographic influences, as are more quantitative approaches to terrain corrections for polygon size and speed analysis.
- More rigorous methods for assigning polygon scores based on quantitative overlap are required. The quantitative spatial analysis methods and approach of Baker and Libarkin (2007), which build on the methods reported here, add significant new formalism to the hotspot generation process using geospatial analysis techniques. Collaboration between many research groups will be required to flesh out this approach and expand sample sizes to ensure robust conclusions.
- How can one use this methodology in real time or near-real time to enhance student learning in field camp settings? Application of these research tools as assessment measures could significantly advance field education in the geosciences.

This chapter contributes new, independent tools for measuring and monitoring field geology performance that do correlate with traditional map grading and which give higher resolution understanding of specific difficulties to geoscience education researchers working with field based learning at this advanced undergraduate level. We have also shown in this chapter that application of insights and guiding frameworks from cognitive science research can be readily applied to geoscience field problem solving. Our methods show how navigation is related to problem solving, and additionally we have made the contribution of new coding schemes that are readily applicable to other field education settings in the geosciences and potential elsewhere. These tools provide useful new methods for understanding the development of this type of expertise, and provide a new avenue for work by other researchers exploring various stages and settings of this poorly understood process.

Broader interest in this chapter is based in the potential practical applications of these techniques to study any area of science education research that considers navigation through space coupled with problem solving. Studies in museum science, classroom dynamics, and other field science applications could potentially benefit from a modification and re-tasking of our approach. This method could also be adapted to real-time monitoring of student problem solving and progress in educational settings.

Discussions with the Marshall research group about this work have also revealed striking similarities with patterns developed in eye tracking research that we have not discussed here. Our methods and approaches with track coding and polygon scoring have already flowed back from this chapter in to that research area, and the unexpected similarity of eye tracking and field navigation patterns with similar problem solving task characteristics but vastly different temporal and spatial scales offers a new comparative window into basic problem solving cognition and the role of expertise.

Implementing Earth Systems Science

It is widely known through a number of national studies that urban students, particularly ethnic and racial minorities, are under-represented populations in the sciences. Increasing recruitment of minority students into the sciences is a nationally Recognised challenge and a national goal. It is also widely known that undergraduate science courses serve as the training ground for the future science workforce, teacher training, and are often the highest level of science exposure for the general public with a college background. For these reasons, the importance of teaching undergraduate science courses has gained national attention. Recent pedagogical emphasis focuses on improving methods of instruction, engaging students-particularly from under represented populations, and the inclusion of systems-oriented Earth science content. However, less attention has been paid to evaluating student responses and performance to curricular changes.

The University of Missouri-Kansas City (UMKC), Department of Geosciences and Environmental Studies Program, with a grant from

NASA's Earth System Science Education 21 Program (ESSE 21), restructured fundamental program curricula. Organising around the concept of Earth system science, four core courses focus on engaging urban students and incorporating urban environmental content. This chapter discusses the curricular changes made to the first of these courses, our introductory level science course, Understanding Earth. Changes consist of an Earth systems orientation, active learning, and an urban place-based foci. Also presented is an evaluation of students performance and response to the restructured curriculum in terms of engaging urban students from underserved populations and non-majors, and measuring student self-efficacy as a means of gauging success.

We live in a time of vastly expanding urban landscapes, growing resource pressures, and increasing societal impacts from natural hazards. Engaging students in the importance and issues Earth science presents today is critical for their future. As citizens they will increasingly need to understand Earth systems concepts in order to make critical economic, political, and social decisions. UMKC, similar to many universities, is a comprehensive urban university serving a diverse population- 24% of the student body comes from traditionally underserved populations. The redesigned course curricula focuses on the interconnection between Earth systems as the means to understanding how the Earth functions. Urban settings offer tremendous and under-utilised opportunities to explore Earth systems science.

The new curricula also deliberately make strong connections to our students' intrinsic urban sense of place. The place-based approach in education is a well established practice and highly effective method of teaching Earth science. While a popular trend in rural and natural settings, employing place-based approaches in urban settings is an untapped resource. Using the urban environment is another opportunity to engage urban minority students in the Earth sciences. The relevance of Earth science is built into the new curriculum by relating Earth systems to the students' sense of place, the human constructed landscape.

Our goals in changing the course curriculum are to:

1) improve the science foundation by using an Earth systems perspective and disseminate this integrated science perspective to a wide range of students: majors, non-majors, non-traditional students, and preservice teachers. Our objective in providing

students with a strong science foundation is to emphasize the dynamic and complex interconnection of processes across the atmospheric, geological, hydrologic, and biological systems.

2) better engage all students in Earth system science, but in particular non-majors and underserved populations from primarily urban backgrounds.
3) provide students the scientific knowledge base to empower their decision making regarding the interconnection of Earth's physical systems and human activities.

The entry level non-major service course, traditionally large-format, is familiar to many science departments. The large format style and non-major demographics present specific challenges to redesigning curriculum. Accomplishing the curriculum restructuring included faculty participation in several teaching workshops, redesigning course content, development of new topical modules, and evaluation tools. While instituting changes in the curriculum, we conducted evaluations at many scales throughout the semester in order to better evaluate the impacts of the changes, and students' response to them. The evaluation tools, both quantitative and qualitative, attempted to capture students' response to and impact of the curriculum changes at several time scales- daily, weekly and semester. Evaluating course mechanics, students' response to Earth systems science, and student performance is critical to determining if we are meeting our objectives. The results are compared to the course statistics for the year prior to the curriculum changes (2003).

Changes in curriculum for the new Understanding Earth course consisted of changes in both structure and content. The curriculum also incorporated new teaching skills and methods for engaging students more directly with the material, their class, and community. The original course structure consisted of three examinations, a few random quizzes and five journal entries about environmental issues. A wider range of assessment methods broadening the evaluation of student knowledge base, critical thinking skills, and student engagement with course material. The curricular changes also required a wider range of evaluations that attempted to capture the impacts of curricular changes both small and large.

Changes to the course curriculum consisted of presenting the interactions of the primary Earth systems; atmospheric, hydrologic,lithologic, and biologic-including human impacts in a wide variety of formats. To get beyond the more familiar and passive approach used in large format courses of Power Point lecture, new teaching strategies incorporated many active learning techniques. These techniques sparked reflective and critical thinking through in-class activities, peer-to-peer discussion, and group work. The course reorganisation specifically consisted of the following:

4) Restructured course content focused on the Earth system science approach to exploring the physical Earth functions. This approach presents the interconnections of physical systems and their critical intersection with human landscapes. Since the majority of the students are from the Kansas City metropolitan area, urban environmental content, applications, and issues were strongly incorporated into all aspects of course content.
5) New teaching approaches incorporated numerous active learning techniques and reduced standard lecture time.
6) Assessment strategies incorporate quantitative and qualitative approaches to evaluating students' Earth systems science content knowledge and integration of concepts on a number of levels. Students' self-efficacy towards Earth science was engaged through a progression of activities beginning with addressing individual learning styles, and continued in individual course activities. Evaluating self-efficacy was one means of measuring overall course impact.

7

Hypothesis Testing and Geoscience Proposal

What first-year undergraduates know about scientific methodology likely comes from a high school science class in which they learned a linear scientific method: identify a problem, make a hypothesis, gather data, and test the hypothesis. Yet geologists commonly follow a nonlinear path towards testing hypotheses, and we tend to work on many hypotheses at a time, rarely fully accepting or rejecting any of them. For undergraduate geology majors (and even graduate students), learning the nuances of testing hypotheses like a practicing geologist is a transition, and few college-level geology textbooks adequately support that transition. For example, the textbook I use to teach Physical Geology, Marshak's Essentials of Earth, articulates a linear version of the scientific method. By way of more detailed explanation, Essentials of Earth carefully distinguishes between a hypothesis and a theory, using as a case study the development of plate tectonic theory from the hypothesis of continental drift. Comparable descriptions of the nature of hypothesis testing, the distinction between a hypothesis and a theory, and the case study of the history of ideas about drift and tectonics occur early in both Monroe, Wicander, and Hazlett's Physical Geology and Chernikoff and Whitney's Geology.

It seems natural for introductory geology textbooks to use plate tectonic theory to illustrate elements of the scientific method. I joke with my Physical Geology class that the story of the plate tectonic revolution is the American geologist's Passover story; we delight in its telling and retelling. The story offers opportunities to elaborate upon any number of subdisciplines in geology, from paleontology to rock magnetism. Wegener makes for a wonderful hero - dead in the field before his time. And the embrace of plate tectonics in America is certainly an excellent case study of the development of a theory: plate tectonics only gained widespread embrace after support developed within many different subdisciplines; it was a hypothesis that has now survived repeated challenges; and it has predictive power. Perhaps introductory geology textbooks focus on this careful explanation of the scientific usage of the word theory because of the ongoing culture wars over evolution (which is, after all, "only a theory"). We hope that our undergraduates become scientifically literate, and literacy involves a clear understanding of what a scientific theory is.

Yet we also hope that our undergraduate geoscience majors become more than literate. We hope that they become skilled researchers and sophisticated thinkers. Specifically, we want them to learn a nuanced and realistic approach to designing hypotheses and developing tests for them. I have found surprisingly little support for this learning in college-level geoscience textbooks. For example, the two texts I use to teach upper-level courses each year, Davis and Reynolds' Structural Geology of Rocks and Regions and Prothero and Dott's Evolution of the Earth, offer next to nothing. Among the textbooks I regularly consult for teaching, the only book to offer detailed and thoughtful description of the way that geologists think about designing and testing hypotheses is Tectonics by Twiss and Moores (in their "Interlude"). Fortunately, more resources are out there in the form of a handful of excellent papers in widely available journals. Below, I discuss five classic papers written for American geologists (and mostly by American geologists) about methodology and the nature of hypothesis testing in the geosciences. Each paper offers an example other than plate tectonic theory, and each is accessible enough to use as supplementary reading in an undergraduate geology class (I hope, too, that this piece is accessible and succinct enough to use as supplementary reading in an undergraduate geology class). Finally, in the conclusions, I share some of my experiences using proposal writing as a way to teach the geoscientific method to undergraduate geology majors.

Hypotheses and Hypothesis Testing

The methodological concept that is most likely a part of the everyday vocabulary of a geologist is "the method of multiple working hypotheses," espoused by the eminent geologist T.C. Chamberlin (1897). Essentially, Chamberlin warns against focusing one's research on testing a single hypothesis at a time, because one is likely to favour it to the detriment of understanding the problem. Instead, he argues for developing multiple hypotheses in the early stages of a study.

Chamberlin uses a family metaphor for his argument: In developing the method of multiple working hypotheses, the effort is to bring up and into view every rational explanation of the phenomenon in hand and to develop every tenable hypothesis relative to its nature, cause or origin, and to give all of these as impartially as possible a working form and due place in the investigation. The investigator thus becomes the parent of a family of hypotheses; and by his parental relations to all is morally forbidden to fasten his affections unduly upon any one.

At its simplest, Chamberlin's essay is an admonition to keep an open mind. His idea of multiple working hypotheses permeates the geologic literature on methodology (some papers refer to it as "The Method"), and it is a touchstone for most other writing on earth science methods. The great investigator is primarily and preeminently the man [sic] who is rich in hypotheses. In the plenitude of his wealth he can spare the weaklings without regret; and having many from which to select, his mind maintains a judicial attitude. The man who can produce but one, cherishes and champions that one as his own, and is blind to its faults. With such men, the testing of alternative hypotheses is accomplished only through controversy. Crucial observations are warped by prejudice, and the triumph of the truth is delayed.

Unlike Chamberlin's essay, in which he outlines reasons for using multiple working hypotheses, Gilbert's essay teaches by example. Gilbert details a field-oriented geomorphology problem and three hypotheses to explain it. The problem he cites is the variable elevation of the paleo-shoreline of ancient Lake Bonneville (a larger, Pleistocene manifestation of Great Salt Lake in Utah). Assuming that the shoreline was level when the lake was full, Gilbert reasons that the shoreline is no longer level because of subsequent crustal warping. Gilbert focuses on the difference between the highest and lowest shoreline elevations. He presents calculations by himself and a colleague, based on assumptions about the physical properties of the earth's crust, on the maximum amount of uplift that could result from each of three hypotheses. Only one of his hypotheses appears to be capable of producing as much uplift as the observed elevation difference, so Gilbert tentatively discards the other two hypotheses and recommends further tests of the survivor.

Gilbert's example is astonishingly current for two reasons. Firstly, rather than validating any one hypothesis (as the linear scientific method would dictate), he selects among his hypotheses by elimination.

In his own words, he casts aside the weaklings. This method of hypothesis testing, also called falsification, responds to the rub that scientists may never really prove that a hypothesis is true, because there is always the possibility that new data will show that it is false. Instead, the only way to stop working with a hypothesis is to show that it is false. Secondly, Gilbert tests his three hypotheses by running numbers in about the same manner as a modern geologist might design a handful of simple computer models. His modelling adds to the strength of the surviving hypothesis. Not only is it the only hypothesis left, it is also reasonable, given the best quantitative knowledge of earth processes and materials. A similar use of modelling to strengthen hypotheses is currently common in the geoscience literature. A 1933 essay on the scientific method by Douglas Johnson offers a different approach to hypothesis testing. Johnson also uses an example of a geomorphic problem: the seaside association of rocky cliffs and erosional benches located above high tide line.

In playing with this example, Johnson stresses the importance of analysis, which he defines as "the process of separating observations, arguments, and conclusions into their constituent parts, tracing each part back to its source and testing its validity, for the purpose of clarifying and perfecting knowledge." In other words, be careful with your assumptions. Johnson encourages multiple hypotheses. To test these hypotheses, he recommends a deductive process of analysing each hypothesis in order to determine its consequences: for example, details that might characterise the benches in the case that the hypothesis is valid. These details are useful for validating or falsifying (i.e. eliminating) individual hypotheses, the basic process demonstrated by Gilbert. Some geologists call this method prediction because one predicts, based on the various hypotheses, what one might observe in the field. Johnson's example contrasts with Gilbert's because his tests involve gathering or reconsidering field data rather than constructing simple models. For example, Johnson selects a hypothesis regarding storm waves, and he conjectures that this hypothesis predicts that the benches have loose debris on them. Is there debris on the benches? Unfortunately, Johnson's example proves more confusing than illuminating at this point because he goes on to invalidate the assumptions underlying the discriminatory character of this prediction. Nonetheless, Johnson makes a useful point:

There will usually be found, however, some one or more consequences peculiar to hypothesis A, while certain others are peculiar to hypothesis B, and so on. It is these unlike consequences which have the highest critical value in discriminating between valid and invalid hypotheses, and it is on these that the investigator will most depend on drawing conclusions.

In other words, thoughtful geologists will gather the data that will allow them to choose between competing hypotheses.

Another excellent essay on hypothesis testing in the geosciences is Carol Cleland's 2001 essay. She argues that geoscientists rarely eliminate hypotheses, for the same reason that Johnson steps away from eliminating his hypothesis about benches: the realisation (or fear) that some of the assumptions underlying the usefulness of the test are invalid. In addition, Cleland argues that geoscience hypotheses are unusually difficult to falsify. Events are too complicated and time frames too long to set up experimental tests. Critical evidence might be lost or beyond our ability to identify. Instead, she posits:

A look at the actual practices of historical researchers, however, reveals that the main emphasis is on finding positive evidence- a smoking gun. A smoking gun is a trace [i.e. data gathered from the rock record] that picks out one of the competing hypotheses as providing a better causal explanation for the currently available traces than the others. Cleland's "smoking gun" idea bears some resemblance to Johnson's "unlike consequence," but Cleland's essay contains a much clearer, more detailed, and more modern exposition of the idea. In addition, she gives an example that works: the cause of extinction of the dinosaurs. High iridium concentrations in sediments deposited at the same time as the extinction are the smoking gun. Because iridium is much more common in space than on Earth, researchers interpret this observation as good evidence for a meteorite impact. The impact and the extinctions occurred at the same time, supporting a cause-and-effect relationship. Cleland argues that field data, rather than laboratory experiments or computer models, provide us with smoking guns:

This brings us to a crucial point: although computer-aided models may suggest what to look for in nature, and traces and some auxiliary assumptions may be investigated in the laboratory, one cannot experimentally test a historical hypothesis per se; to recapitulate, the time frame is too long and the test conditions too complex to be replicated in the lab.

Certainly, most geoscience research starts with field observations, which lead to questions based on these observations. Ultimately, hypotheses must agree with field data. This necessity is one of Johnson's central points. Cleland is arguing, however, that only field data allow the geoscientist to select one hypothesis among many, if the hypothesis concerns the causes or nature of past events. In contrast, Gilbert's example from Lake Bonneville uses simple calculations, quite like computer-aided modelling, to falsify two of his hypotheses, allowing him to tentatively select the only remaining hypothesis. These two approaches differ significantly, and both are useful. Even if experimental or model results are not smoking guns, they allow geoscientists to explore which hypotheses are possible, given the current state of quantitative knowledge of the pertinent physical processes. Ultimately, our knowledge of what is possible is constantly evolving. As a result, models and experiments do not provide a final test. Nor do field data. For example, some geologists continue to favour different causes for the extinction of the dinosaurs, even though the field data are clear that a giant meteorite impact occurred at that time. Perhaps the meteorite didn't cause the extinction, or perhaps the story is more complicated. Why did mammals survive? The geoscience community is diverse in its thinking, and geologic events and phenomena are complex. Can we ever consider any specific geologic hypotheses to be fully tested? Having posed this question, it is illuminating to turn again to Chamberlin's 1897 essay.

Complexity and Circular Reasoning

Although Chamberlin's essay focuses mainly on arguments for using multiple working hypotheses, he also provides the example of the origin of the Great Lakes: The mooted question of the origin of the Great Lake basins may serve as an illustration. Several hypotheses have been urged by as many students of the problem of the cause of these great excavations. All of these have been pressed with great force and with an admirable array of facts. Up to a certain point we are compelled to go with each advocate.

It is practically demonstrable that these basins were river valleys antecedent to the glacial incursion. It is equally demonstrable that there was a blocking up of outlets. We must conclude then that the present basins owe their origin in part to the preexistence of river valleys and to the blocking up of outlets by drift. That there is a temptation to rest here, the history of the question shows. But on the other hand it is demonstrable that these basins were occupied by great lobes of ice and were important channels of glacial movement. The leeward drift shows much material derived from their bottoms. We cannot therefore refuse assent to the doctrine that basins owe something to glacial excavation. Still again it has been urged that the earth's crust beneath these basins has flexed downward by the weight of the ice load and contracted by its low temperature and the basins owe something to crustal deformation. This third cause tallies with certain features not readily explained by the others. And still it is doubtful whether all of these combined constitute an adequate explanation of the phenomena. Certain it is, at least, that the measure of participation of each must be determined before a satisfactory elucidation can be reached. The full solution therefore involves not only the recognition of multiple participation but an estimate of the measure and mode of each participation. For this the simultaneous use of a full staff of working hypotheses is demanded.

Chamberlin's example is surprising because it gives no sense of hypothesis testing. He presents multiple hypotheses, but he does not eliminate any hypotheses, nor does he shore up support for others with a smoking gun, an experiment, or a realistic quantitative calculation. Rather than appeal to any further testing of these hypotheses, Chamberlin indicates that future research might weigh their relative roles or identify even more hypotheses to add to the list of factors that contributed to the development of the Great Lakes. Chamberlin is fully aware of the significance of his example. The final paragraph of his essay begins: "the studies of the geologist are peculiarly complex. It is rare that his problem is a simple unitary phenomenon explicable by a single simple cause."

Chamberlin's articulation of the complexity of geoscience explanations rings true. But none of the essays discussed above fully addresses the significance of this complexity for hypothesis testing. A fascinating 1995 essay by Robert Frodeman fleshes out some of these concerns from a different perspective. Rather than understanding hypothesis testing as the only goal of research, Frodeman understands it as one of the many tools that geologists use to understand past events. Another tool, for example, is analogies between modern and ancient processes and events. As a description of the way geologists reason, Frodeman uses an example of a geologist looking at an outcrop:

More to the point, our understanding of an outcrop is based on our understanding of the individual beds, which are in turn made sense of in terms of their relationship to the entire outcrop. This back-and-forth process of reasoning operates at all levels; wholes at one level of analysis become parts at another. Thus, our understanding of a region is based on our interpretation of the individual outcrops in that region, and vice versa; and our interpretation of an individual bed within an outcrop is based on our understanding of the sediments and structures that make up that bed, and vice versa. On a still more complex level, our overall comprehension of the Cenomanian-Turonian boundary event is determined through an intricate weighing of the various types of evidence (e.g. lithology, macro- and micropaleontology, and geochemistry). This overall interpretation is then used to evaluate the status of the individual pieces of evidence. Such circular reasoning is usually viewed as a vice, a logical fallacy to be avoided at all costs. But Heidegger argued that this type of circularity is not only unavoidable, it is actually, if properly handled, the means by which understanding progresses.

Anyone who has inspected an outcrop or participated in a panel discussion with a group of geologists can see that Frodeman's description is accurate. This circularity is different from the more linear process we invoke when we talk about hypothesis testing. In addition, it is easy to see how both field data and experiments or quantitative models fit into the circular reasoning process. The ease of this fit is clear in the geologic literature, where geologists often present both new data and model results in the same publications.

What are the implications of Frodeman's description for the role of hypothesis testing in the geoscientific method? Do geologists even test hypotheses? Yes, hypothesis testing is an intuitive and ubiquitous way of thinking and asking questions, even though it is not the only way that research progresses. Hypothesis testing is a fundamental part of our common methodology, even though this common methodology encompasses the diversity of approaches outlined above.

Ideas for Teaching

When I first started to think about how I might teach undergraduates a more nuanced and realistic sense of how working geoscientists articulate and test hypotheses, I decided to start assigning proposals as writing assignments in my Historical Geology class. Proposal writing requires a very clear statement of hypotheses, an understanding of how to test them, and a set of arguments showing that the hypotheses are worth testing. I wanted to give students experience writing a few different proposals over the course of a semester so that they could practice the format. Focusing these writing assignments was difficult for me at first because most undergraduate proposal-writing experience relates to independent research projects outside of the classroom (often a proposal for senior thesis research). Furthermore, most undergraduates, and even many graduate students, have not seen written arguments involving hypothesis testing because they have never read a research proposal. Proposals are almost always confidential; usually the only people who read them are the handful of scientists who review them. How could I incorporate proposals into coursework? At first, I asked students to pick examples of controversies from their text and write proposals about competing hypotheses and ideas for how to move the thinking forward.

Most students worked with the controversies that were best articulated in the text: causes of the Ordovician trilobite and the Pleistocene megafauna extinctions. After two years of reading papers on these two debates, I became frustrated with my own assignment. By reviewing stalled debates, my students may have been learning how to articulate scientific questions as hypotheses, but their papers were repetitive and relied too heavily on the text. Over the past two years, I have used recent research in the journals Science and Nature as springboards for student proposals. Each year, I make a list of the past year's historical geology papers in these two journals. Each student picks two papers from the list and presents them to the class. Presentations occur each week, and other students can use any presentation (and associated paper) as a starting point for a written proposal to move the research forward. Students have a lot of choice: they choose which papers to present, and they choose which papers to explore more fully in their proposals. Consistently, the most significant challenge of the proposal is finding an appropriate scale for hypotheses-somewhere between testing the hypothesis of uniformitarianism and testing the hypothesis that someone might eventually find a more complete fossil of a poorly understood species. I remind students that most scientific research occurs in small steps, and many researchers start new projects by first considering what kinds of data they are able to gather or model.

Then they consider what questions these data or models might address. Articulating and contextualising these questions as hypotheses is often the last step in putting together a proposal. Even though writing proposals is challenging, students generally like the assignment: it brings cutting edge research into the classroom each week, it allows students to write about the topics and techniques that most interest them, and it encourages original and critical thinking. What's not to like about that?

Improving Spatial Visualisation Skills

Three-dimensional spatial visualisation is an extremely important skill in many fields involving science, technology, engineering, and mathematics, including the geosciences. For example, geologic structures like faults and folds are inherently three-dimensional features; successful students must be able to visualise how these structures interact with topography, and extend to depth in the subsurface. Despite the importance of spatial visualisation in many fields, it is rarely taught thoughtfully or tested for in K-12 or undergraduate curricula. Consequently, many talented students whose spatial visualisation skills are insufficient to accomplish course assignments do poorly in visualisation-intensive classes and switch their field of study away from these disciplines. This process results in an unfortunate homogenisation of thinking styles among students in these disciplines and may result in a lack of novel ideas and problem-solving strategies in fields that rely heavily upon innovation.

Some students are naturally better at spatial visualisation than others. This range of natural ability often leads to frustration in the class for both students and instructors. Students whose spatial skills are naturally inferior to others' often feel frustration at being unable to complete tasks that are easy for others. Similarly, an instructor must interact with a diverse group of students whose different abilities may not allow them all to have the same opportunity to perform well on assignments and exams, through no lack of effort or interest on the students' part. Research has shown, however, that people's spatial visualisation skills can often improve with practice.

The main purpose of this project was to test whether students' three-dimensional visualisation skills can be improved through practice provided in geology courses. We used two different strategies on two different student populations to answer this question. We present the results of each study below, following a discussion of spatial visualisation.

Spatial Visualisation

Spatial visualisation is a complex process that involves both visual abilities and the formation of mental images. Because of the importance of spatial visualisation across many disciplines, it has been studied by a wide variety of workers in science, education, and cognitive psychology. Various classification schemes exist for spatial visualisation, many of which divide the process of visualisation into anywhere from three to ten component skills. Often these component skills overlap, making the terminology confusing.

We use a simple division of spatial visualisation into three component skills useful in a geological context: spatial relations, spatial manipulation, and visual penetrative ability. Spatial relations are the ability to mentally rotate an object about its centre. Rotations may be about one or more axes and can reflect a variety of geological problems on all scales, from crustal block rotations to plotting fault orientation data on a stereonet. Spatial manipulation is the ability to mentally manipulate an image into another arrangement. Spatial manipulation skills are useful in structural geology when envisioning how bodies of rock deform through time, such as migration of a magma body through the crust, or progressive folding of a stratigraphie succession. Visual penetrative ability is the ability to mentally imagine what is inside of a solid object. This skill is less commonly included in many visualisation classifications, but is critical for geoscientists who commonly use cross sections, thin sections, roadcuts, and many other slices through Earth and Earth materials to interpret geologic processes and histories.

Numerous studies have shown that practice tends to improve all students' spatial visualisation skills regardless of their initial ability. Participation in courses with occasional exposure to spatial exercises, including geology seems to improve spatial visualisation skills. More directed interventions using software, handheld objects, and mental imagery practice also improve students' skills. Geologically relevant examples include a study by Lord (1985), who found that students given 30 minutes of weekly practice mentally bisecting three-dimensional geometric figures significantly improved their spatial visualisation abilities, including their spatial relations and visual penetrative ability component skills. Duesbury and O'Neil (1996) used computer-generated two and three-dimensional images to test whether these images aided three dimensional visualisation ability. Practice was given to two groups, one of which allowed for rotation of the images. The authors found that the rotational group performed significantly better on measures of spatial ability. In a controlled study, Piburn et al. (2001) showed that students who received computer-based instructional modules about three-dimensional topics related to geology were better able to solve spatial visualisation problems. Other examples of the benefits of practice with spatial visualisation skills exist for fields including chemistry, mathematics, and geography.

There is active debate in the literature whether gender plays a role in spatial visualisation ability. Some studies, typically of abstract spatial tasks, have shown that males tend to outperform females on spatial tasks but these differences may be due to environmental conditions. Other studies suggest that females have better abilities on certain component spatial tasks. In geology, Piburn et al. (2001) suggest that gender differences can be eliminated by allowing ample opportunities for practice. Although our study does not directly focus on determining if there are gender differences in spatial abilities of geology students, some of our data allow us to comment on this topic.

We present results from two separate studies: the first from Carleton College and the second from the University of Wisconsin. The Carleton College study used pre and post-course surveys from introductory, intermediate, and upper-level geology courses to establish a baseline understanding of students' visualisation skills and how their skills change after enrollment in geology courses. These survey data allow exploration of the differences in the student population among men and women as well as majors and non-majors. The University of Wisconsin study was a qualitative study on a small number of students taking Structural Geology at the University of Wisconsin. Students were asked to solve geologically relevant problems that required spatial visualisation skills and several assessment strategies were employed to test whether students' skill improved. Both studies were designed to test whether visualisation practice provided in geology courses could improve students' visualisation skills.

The Carleton College study is presented first because this dataset provides a picture of students' abilities in a range of geology courses. Chronologically, however, this chapter took place after our University of Wisconsin study. Although this second study is qualitative - the small class size prohibited reliable quantitative analysis - the Wisconsin dataset provides insight about the types of problems students have when solving geologically relevant problems (as opposed to abstract spatial problems) and represents a model for how geoscience educators might test their own instructional materials. We therefore present both studies (out of chronological order) relying primarily on the quantitative results but noting pertinent aspects of the qualitative study when they are especially informative.

The first study we describe explores students' abilities to solve abstract visualisation problems. We developed an abstract visualisation survey and administered this as a pre- and post-course survey to students in Titus' courses at Carleton College. This survey-based design allows us to examine differences across the student population and gain more quantitative information about students' spatial skills. The disadvantage, however, is that we do not know how performance on abstract visualisation tasks transfers to the ability to solve applied geology problems requiring spatial skills.

The survey instrument is divided into three sections, each of which tests a particular component skill: spatial relations, spatial manipulation, and visual penetrative ability. The first section consists of ten three-dimensional rotation exercises from a longer and widely used spatial test for mental rotation published by Guay (1976). (In an earlier version of the survey administered at the University of Wisconsin, we used a two-dimensional rotation task from Ekstrom et al. (1976) that was too easy for college students.) The second section consists of four spatial manipulation exercises published by Ekstrom et al. (1976). The third section consists of fifteen penetrative thinking exercises we created in addition to some published by Crawford and Burnham (1946) and Myers (1953). We designed the survey to minimise the amount of class time necessary, thus each section lasts three minutes and the total time for administering the test is no more than fifteen minutes. Most students do not finish each section although some students are able to finish individual sections.

This survey was administered to Titus' students at Carleton College on the first and last days of three separate courses - Introductory Geology, intermediatelevel Tectonics, and upper-level Structural Geology - taught from Fall 2006 through Fall 2008. AU courses were taught in the mornings and each had a four-hour afternoon lab associated with the course. Data from the same courses from different years were combined to improve statistics: two years of Introductory Geology, three years of Tectonics, and two years of Structural Geology. Specific exercises (i.e. the skill puzzles described in a later section) were used in each course to allow students to practice with spatially-intensive problems. The nature of course material in Tectonics and Structural Geology required spatial problem solving more often than in Introductory Geology. We also compare survey scores in Introductory Geology with final course grades. For each of the three tasks in the survey, there is marked improvement between the pre- and postcourse survey, which is always statistically significant with $? < 0.05$. Between courses, there is also general improvement in scores, where higher percentages are observed in higher-level courses. These intermediate- and upper-level courses more consistently require visualisation skills to understand course content and are typically only taken by majors or potential majors.

Gender Comparisons

The results are also divided by student gender. In Introductory Geology, men outscore women on all three tasks on the pre- and post-course surveys. However, differences between genders are not statistically significant and women typically score above the pre-course survey levels of their male counterparts on the post-course survey. In Tectonics and Structural Geology, the pattern is similar but these courses attract students who are considering or are geology majors. Students' overall scores are higher than those in the Introductory Geology course and statistically significant gender differences are observed only for the spatial relations (mental rotation) task. In contrast, there are no differences between gender for the spatial manipulation or visual penetrative ability tasks.

Similar gender differences can also be observed, which shows results compiled from the 28 students who have taken the visualisation survey four times as a result of taking two of the three possible courses. These students are all likely geology majors. The gains between the first and fourth exams are all significant with $p < 0.01$. Perhaps more importantly, the differences between men and women are completely absent from the spatial manipulation and visual penetrative ability tasks. In fact, women outscore men on three of the four surveys for visual penetrative ability. The differences for the spatial rotation task, however, are still apparent (although not statistically significant given this smaller sample size).

Majors versus Non-majors

From the pool of students who have taken Introductory Geology, we compared those who have not yet taken another geology class (and probably will not) with those who have taken or are signed up for a second geology course. Most of these differences are not statistically significant in part because of the smaller sample size of the potential major group. However, our small dataset suggests that potential majors outscore non-majors on both pre- and post-course surveys, that potential majors have increased gains between surveys, and that non-majors do not reach the pre-course survey level of their potential major counterparts.

Although there is significant scatter in this dataset, students receiving higher grades tend to score higher on the three abstract visualisation tasks. This interpretation is supported in two distinct ways. First, students who earned a B+ or above had scores > 10% on the post-survey for each of the three tasks. Second, students who earned Cs had lower visualisation scores in general (typically not scoring > 60% on any section) and those two students who failed the course had particularly low scores (< 50%). It is worth nothing, however, that very low scores on any of the three tasks is not an absolute predictor of poor course performance as several students who received grades as high as B scored very poorly (< 10%) on one or more of the abstract visualisation tasks.

Qualitative Study

Over several semesters, we developed instructional materials to provide students with practice using spatial visualisation skills in introductory and upper-level geosciences courses at the University of Wisconsin. Here we describe in detail an implementation of those materials in a 2005 Structural Geology course. Structural Geology is a highly visual discipline that aims to observe and understand the processes and implications of rock deformation. Developing a thorough understanding of rock deformation often involves mentally projecting, rotating, and generally manipulating spatial data. The fourteen students enrolled in this course were all upperclass majors in geology or geological engineering.

Our instructional materials were implemented in two ways: (1) data visualisation exercises were given in a controlled study in the laboratory portion of the class (the seven students in each lab section represented the experimental and control group) and (2) geologically relevant exercises involving component visualisation skills were given to all students in the lecture portion of class. To evaluate the effectiveness of our materials, we compare answers on exams from the experimental and control groups as well as a voluntary survey six months after the end of the course. Even though quantitative assessment of our instructional materials is not possible given the small number of students enrolled in the course, the results of this experiment are promising. Further, we offer our approach as a model for geoscience educators interested in assessing the effectiveness of particular teaching strategies or materials (and conclude with ideas on what types of pitfalls could be avoided by careful experimental design).

The first set of instructional materials for this study at the University of Wisconsin was designed to provide practice with stereographic projection, a common visualisation tool in Structural Geology. In our experience, stereographic projections are often difficult for students. Consequently, we hypothesized that if stereographic projections could be mastered, they might serve as a gateway into more complicated visualisation problems.

There were two laboratory sections in this course, which met at the same time of day on consecutive days and were taught by the same graduate student TA (either Horsman or Titus) in any given week. We conducted a small-scale controlled experiment in which the experimental section received stereonet-related problems while the control section did not. Possible answers were discussed immediately following the exercise and the instructor offered insight into how an expert might solve each problem. For the experimental group, 10-15 minutes of each lab session were devoted to solving and discussing a stereonet-related exercise.

Our exercises emphasize practical spatial visualisation skills in geology including: (1) extracting and/ or plotting three-dimensional data on a stereonet, (2) Recognising the difference between linear and planar data, (3) estimating where data would plot on a stereonet without using technical methods, (4) describing and/ or generalising stereonet data in words, (5) rotation of stereonet data about different axes, (6) application of stereonets to straightforward geologic problems on a variety of spatial scales, and (7) application of stereonets to more abstract and complicated geologic problems. However, Recognising that students have a variety of preferred learning styles, there were several different types of activities to address different learning styles including written independent assignments, oral descriptions of problems, and group activities with tactile objects.

We used student performance on exams to assess the effectiveness of this mini controlled experiment. Students had three exams during the semester; for each, we designed questions to specifically test spatial visualisation skills, including a stereonet exercise within a multi-step visualisation question.

On the first two exams, the experimental group outperformed the control group on visualisation problems (especially after the removal of one outlier student who performed particularly poorly in all aspects of this course). For the non-visualisation questions, there was no difference between the two groups of students on these two exams. For the third exam, the difference between groups disappeared. However, for this exam, students were told that if their score was better than their three-exam average, it would count as their exam score for the entire course. This may have encouraged students to study more diligently for Exam 3 than for earlier exams. Further, it is interesting to note that for the third exam all students did better on the visualisation questions than they did on nonvisualisation questions. This may be linked to the lecture-based activities that all students received described in the following section.

To give students practice developing their visualisation skills within a geological context, we also developed a set of formal assessment tools that we call skill puzzles for use in the classroom part of Structural Geology (for all fourteen students). The rationale for these exercises came from a pilot study in the previous Structural Geology course in 2004, where students were asked to solve problems that required spatial skills, such as making structure contour maps or drawing crosssections from block diagrams similar to those used by Kali and Orion (1996). These spatial topics had been covered (at what we thought was extensive detail) in a prior course. However, we found that most students were unable to solve what were simple problems for the professor and TAs. The idea of skill puzzles grew out of our pilot exercises and was designed to provide more practice solving spatial problems with content pertinent to course topics.

Implementation

Students solved (anonymously) one skill puzzle per week during lecture. Each skill puzzle required some visualisation skill to solve a geological problem relevant to the lecture material. Typical topics covered included: (1) interpreting topographic and geologic maps, (2) making geologic maps and cross-sections, (3) making and interpreting structure contour maps, and (4) stress and strain analysis.

Immediately following the exercise, the instructor and students engaged in a discussion of possible answers to the skill puzzle. This discussion included a description of how an expert might solve the problem. The total time devoted to both problem solving and discussion was usually 10 minutes. However, on at least one occasion, the discussion ended up generating questions that lasted considerably longer, revealing areas where more formal instruction was clearly necessary.

Effectiveness

Skill puzzles are first and foremost useful for diagnosing whether particular students have trouble with particular visualisation component skills. We show example answers to a cross-section-based skill puzzle, where visual penetrative ability is required. About half of the students in Structural Geology solved this problem correctly (and their answers are not shown). The other half had a variety of visualisation problems ranging from (i) not Recognising the fault to (ii & iii) not understanding that the fault must cut through the entire cross-section. If the fault was Recognised, it was (iv & v) curved instead of planar or (vi) the student did not realise that the rocks below the fault were also folded. Interestingly, those who correctly identified the fault (the bottom row) but incorrectly identified its geometry also felt compelled to label their fault with text. In contrast, those students who answered correctly did not use text in their solutions.

We also developed scoring rubrics (from 0 to 5 points) for each skill puzzle based on the type of question and component skills needed to answer the question. To examine whether our mini-controlled experiment affected skill puzzle scores, the average scores for students in the experimental and control groups. The experimental group typically outperformed the control group but this difference is not statistically significant. In fact, the variety of puzzle types, changing difficulty levels, and changing student attendance rates make it difficult to demonstrate any significant trends in students' abilities (experimental or control group) to solve skill puzzles.

A more concrete measure of improvement can be seen in the results from giving students the same skill puzzle twice: once at the start of the semester and again as a bonus question on the final exam. In this example, there was a marked improvement in students' ability to determine the general shape and orientation of the structure contours for the folded layer - from an average of about 48% correct to 72% correct. We cannot specifically determine whether this improvement is due to real gain or whether students remembered the in-class discussion of the answer. However, because several answers on the final exam were actually more correct than the solution shown in class (because their structure contour lines continue across the entire map area), we suspect that most students actually developed a deeper understanding on this type of visualisation-intensive problem.

Post-course Survey

Six months after the end of the course we sent students several visualisation questions and asked them to return the completed problems to us. The problems included visualisation-related course material and tested skills like plotting information on stereonets. The intention of this survey was to examine long-term retention of visualisation skills in the class as a whole and also to look for differences in retention between the test and control groups.

The sample size was especially small for the postcourse follow-up exercises. We received completed exercises from 7 of the 12 students to whom we sent the voluntary work (4 students from the test group and 3 from the control group). Despite the small number of responses, some trends are apparent in the students' answers. All students retained basic stereonet plotting skills and were able to make accurate meaningful interpretations of basic structural information (e.g. fold orientation analysis) recorded on stereonets. Students' retention of more complicated structural analysis techniques (e.g. interpreting stress/strain orientations) was more limited - some students were still proficient and others were not. Both retention and performance by test group students appears to be superior to those of control group students, although the small sample size precludes statistical verification.

While our results are qualitative, they suggest that practice can improve students' spatial skills, which itself is a promising finding for geoscience educators. Based on the controlled experiment, students in the experimental group received more practice and they also typically outscored the control group on exams and skill puzzles. Based on the lecture exercises, all students tended to do better on visualisation problems by the third exam than they did on nonvisualisation problems. Thus, a limited amount of time devoted to practice and discussion of visualisation problems in class (as little as 10 minutes per week) may significantly affect students' spatial skills.

Iterative development of these materials and methods, in addition to resources about classroom assessment taught us a great deal about the scholarship of teaching. What we discuss here, however, are several factors that we wished we had known at the start of this experiment. First, assessment of materials should not only occur in exam situations. Many students find exams to be stressful, a condition that can affect their ability to perform tasks that would otherwise be straightforward for them. For this reason, exam results should be used along with other, non-high-stakes tools to evaluate student performance. Second, this independent measure of student improvement might include a non-graded assessment administered as a pre- and post-course survey (such as the one used in the Carleton College study). In our University of Wisconsin study, we incorporated this into our study design by giving an abstract visualisation test with questions. However, some sections of our test were too easy for college students and everyone answered them correctly. We also gave unlimited time on this survey, which removed differences between students who could solve problems with ease (and therefore likely have better developed visualisation skills) with those who used up to ten times as long to solve the same problems. Third, because our controlled experiment provided more time-on -task for students in the experimental group, we cannot determine whether their gains are due to this additional (yet small amount of) time and not necessarily to the instructional materials themselves. In a subsequent, more nuanced investigation we provided a control group with abstract visualisation skill puzzles while the experimental group worked on skill puzzles with geologic context. This dataset, like the one presented in more detail here, also suffered from small sample sizes and the results are therefore not presented here. Last, collecting data over several years would help alleviate the problems of small sample sizes for those interested in assessing materials at institutions with small course enrollments.

The Carleton College data provide us with a better understanding of the range of students' abilities at a variety of points within a geology curriculum. In aggregate, students in Introductory Geology have less well-developed spatial visualisation skills than those in higher-level courses. Those whose natural ability is high at the start of Introductory Geology are likely to be among the pool of students expected to continue in geology. Those who perform extremely poorly on the pre-course survey might be singled out for extra attention throughout the course to improve their chances of understanding course material and passing the class. In courses with both potential majors and majors (introductory and intermediate-level courses), visualisation skills are better developed among all students and gender difference are only apparent for the spatial relations (mental rotation) task, consistent with a meta-analysis of spatial visualisation skills by Linn and Peterson (1985). These results suggest that mental rotation should not be used as the only diagnosis for success at visualisation abilities additional component visualisation skills must also be assessed.

The instructional materials developed in our Unviersity of Wisconsin study are useful for diagnosing which students have particularly problems with geologically relevant visualisation skills. Our results also suggest that a small amount of time devoted to visualisation practice can improve students' skills. We observed this in two ways. First, the superior exam performance by students in the experimental group on visualisation-related questions suggests that the extra practice they received solving stereonet-related problems was beneficial. This observation is strengthened by the similar scores on non-visualisation problems observed for the experimental and control groups. Second, the overall improvement of all students by the third exam on the visualisation questions suggests that the frequent practice in the lecture portion of class was also beneficial. This is a reassuring result, suggesting that differences among individuals are not static and that major gains can be made for students with poorly developed spatial skills.

We recommend that instructors in highly visual disciplines use a pre-course assessment to determine the baseline spatial visualisation abilities of their students. This assessment would allow educators to identify students who would benefit from extra attention devoted to developing their three-dimensional thinking skills, and students who are more likely to succeed in geoscience courses and become majors. This assessment should not be limited to a single component task (e.g. mental rotation) as this may result in a misleading, incomplete picture of students' (in particular women's) spatial visualisation skills. If spatial skills are important for understanding course material, we also recommend that a small amount of class time be devoted to solving courserelevant visualisation problems, as this can improve all students' spatial visualisation skills.

In a quantitative study at Carleton College, surveys of abstract visualisation skills demonstrate that there are differences in skill levels between students who are potential majors and those who are not, suggesting that students may self -select geology as a major based on their visualisation abilities before significant course work. Students in introductory geology courses with very poor spatial skills are more likely to receive poor final course grades than those who have better developed spatial skills. Students' visualisation skills improve in upper-level classes that require more frequent and complex spatial visualisation tasks. Differences between men and women decrease in these upper-level courses as well, except for tasks requiring spatial relations (mental rotation) where men consistently outperform women.

In a qualitative study at the University of Wisconsin, we tested whether students' spatial visualisation skills improved after frequent exposure to a variety of geological visualisation exercises. Our instructional materials included skill puzzles given to all students in lecture and stereographic projection exercises implemented as a controlled study in lab. Comparison of exam scores between the experimental and control groups suggests that students given extra visualisation practice out-performed other students on visualisation questions but not on overall exam scores. These instructional materials are also useful for diagnosing when students have trouble with particular visualisation skills. Considered together, our results suggest that spatial visualisation skills can be improved by participation in a geology course and by frequent opportunities for visualisation practice lasting only 5-10 minutes per week.

Geoinformatics for Earth System Science

The term geoinformatics is used globally with varying connotations, though it has its origin and popular use in Europe. According to the International Institute of Geo-Information Science and Earth Observations (ITC), the Netherlands, technologies supporting the processes of acquiring, analysing, and visualising spatial data form the core of geoinformatics (ITC, 2001). Other researchers consider geoinformatics as a new discipline integrating elements of various disciplines that deal with geographic or geospatial data, i.e. data linked to some locality on the surface of the Earth. These disciplines include, but are not limited to, surveying, remote sensing, geographic information systems (GIS), cartography, geosciences, informatics, computer science, statistics, and management. In recent years the term geoinformatics became popular in the US Earth Science community when the Geoscience Network (GEON), a National Science Foundation funded program, adopted the term to describe a variety of efforts that promoted collaboration between computer science and the geosciences to solve complex scientific questions using advanced information technologies and integrated analysis. I define geoinformatics as the collection, integration, management, analysis, and presentation of geospatial data, models and knowledge that support disciplinary, multidisciplinary, interdisciplinary and transdisciplinary research and education.

The main objectives of this chapter are to (1) demonstrate how geoinformatics can play an important role in the evolution of the systems approach to understanding the Earth, (2) share witn a wider education community the author's experience in creating and teaching an interdisciplinary upper undergraduate level geoinformatics course at the University of Alaska Fairbanks (UAF), and (3) recommend when and where components of geoinformatics should be introduced into the formal educational curricula.

Link Between Earth System Science and Geoinformatics

The very nature of Earth system science and geoinformatics ties them together in a synergistic bond. The all-encompassing nature of Earth system science has been well documented by several agencies and researchers. Though the progression from disciplinary research to multidisciplinary and interdisciplinary research occurred in the middle of the twentieth century, the vision of a holistic approach to the study of Earth as a system began to take on a more definitive shape when global data sets from satellites became available to the community. The space-based observations which formed the basis of the emerging field of remote sensing provided a synoptic and global overview of land, water, and air, facilitating an interdisciplinary approach to studying the Earth.

Study of the Earth as an integrated system progressed in parallel with the developments in satellite and computer technologies. The first Landsat satellite (originally called the Earth Resources Technology Satellite) was launched in 1972, and was so successful that it literally changed the way we looked at our planet (NASA 2005). In the last four decades several thousands of man-made satellites have provided enormous volumes of data (NASA 1997). Powerful computers have made possible the use of software tools (models) which use these observed data to simulate processes operating on Earth and to predict future scenarios. Earth system science relies heavily on archived data and near-real-time data, as well as on modelled data.

In essence, geoinformatics provides the means to integrate the observational data, modelled data, and models that form the building blocks for scientific study of the complex Earth system. An upper undergraduate course in geoinformatics was introduced at the UAF in the spring of 2004. The course is based in the Department of Geology and Geophysics. Though the University does not have an established program in Earth system science, introduction of the course is an effort to move education from a strict disciplinary to a more interdisciplinary realm, and to show a recognition and appreciation for the need to offer educational courses that cut across programs.

How the interdisciplinary course was established Faculty surveys of university environments reveal that research domains are becoming increasingly interdisciplinary and the need to keep up with new data and tools to address research questions is growing, while faculty time and resources are increasingly limited. To overcome these challenges, it makes good sense to pool resources and establish interdisciplinary courses that meet the needs of several existing programs within a university system.

When the Universities Space Research Association (USRA) announced an opportunity for funding new interdisciplinary courses through its Earth System Science Education for the 21st Century program, UAF faculty members from the College of Natural Science and Mathematics, College of Engineering and Mines, School of Education, Institute of Arctic Biology, Geophysical Research Institute, and the Arctic Region Super Computing Centre decided to get together and create a new course in geoinformatics. They envisioned that this course would eventually become a capstone course that would meet the needs of several existing programs at UAF. They believed that ideally the course should be developed with input from both College of Liberal Arts and School of Management faculty. However, the time and complexity of establishing a new university course appeared to be commensurate with the number of departments, deans, and directors involved in the approval process. The faculty team had limited experience in creating new courses and decided to experiment with introducing the course after receiving input and approval from four programs only.

Designing the course involved setting the educational goal and objectives for the course, outlining the course content, compiling lecture and lab material based on the specific objectives, and devising appropriate tools for course evaluation and assessment. In the first two years of its delivery, the course was successfully completed by 16 students from the disciplines of geology, geophysics, engineering, fisheries and ocean sciences, biology, forestry, education, and ecology, and by non-degree-seeking students who were interested in honing professional skills required to work in local industry. The diversity in student profiles was in itself very encouraging and signified the potential of this course to serve as an interdisciplinary capstone course.

The 300 level, three credit, upper undergraduate course is designed for both science and non-science students. It includes classroom lectures and hands-on experience in using data and tools in the computer lab. The lectures lay the foundation for understanding the theoretical aspects of geoinformatics and the labs provide an opportunity for the students to use geospatial data, software tools, models, and visualisation environments. The semester long course has approximately 28 lectures of one hour duration (2 lectures per week), and 14 three hour long labs (one lab per week). The course entails an independent student project where the student applies the knowledge and skills learned in the course to address an Earth system science question of his/her interest.

The overall aim of the course is to develop students' understanding, approach and technical skills in geospatial data acquisition, processing, analysis, management and visualisation for addressing scientific questions related to Earth system science. Specific educational objectives outlined state that by the end of the course, students will be able to

1. Know where to look for and find data, how to download it, and bring it to a useful format for their research or application.
2. Interpret aerial photos and satellite images using elements of photo-interpretation such as tone, texture, shape, size, pattern and association.
3. Use the basic functionalities of an Excel spreadsheet
4. Acquire GPS measurements using portable hand held GPS devices synchronised with Axim palm-tops
5. Integrate GPS data with other GIS data
6. Plan a GIS project
7. Build a GIS geodatabase for an application of their interest
8. Analyse the geospatial data using available software packages (such as ArcGIS)
9. Generate a good map using principles of cartography
10. Explore the role and potential of geoinformatics in research and industry
11. Demonstrate his/her understanding of the subject by completing a small independent project.

One of the challenges in designing the course was to define and structure the course content to meet the ambitious educational objectives. The course itself consists of distinct but interrelated units, each unit addressing a specific aspect of geoinformatics. It starts with an introduction to the systems approach followed by units covering the concepts, techniques, and applications of field data capture, photogrammetry and remote sensing, GIS, database structuring and management, cartography, and geo-information visualisation. A brief description of each of these units is presented here.

Introduction to Earth System Science and geoinformatics (first two weeks) - The first two weeks of the course are used to introduce the concept that Earth is as a dynamic system and processes operating in this system involve complex interactions of the atmosphere, geosphere, hydrosphere (including cryosphere), and biosphere (including human beings). The changes that we see are a manifestation of natural processes and human interactions. The first part of the course is also used to introduce the concepts of geoinformatics.

Two specific examples are used to demonstrate the application of geoinformatics to study the changing Earth. The first is the Digital Tectonic Activity Map that uses geoinformatics to integrate current global datasets of seismicity, volcanism, plate motions, bathymetry, topography, satellite images, and field studies to represent global tectonism and to show how the Earth's surface is changing primarily due to natural causes. The second example used is of a standalone geoinformatics system caned MINEDEMON that integrates an extensive database on anti-personnel landmines in Angola (Africa) and exemplifies how land-cover and land-use patterns change as a result of anthropogenic activities.

Field Data Capture (two weeks) - Field work is an important component of any Earth system science program. Field information constitutes the primary data input for all mapping projects. It is important that these data be recorded in a structured and complete way, and be readily retrievable, not only by the person who recorded the information, but also by others who may want to continue a mapping project later. In this unit students first go through lectures on global positioning system (GPS) and field data collection. They then gain hands-on experience on data acquisition using recreational mode GPS receivers synchronised with handheld Axim notebooks (palm-top computers). A combined GPS-Axim unit costs about $500 making it an affordable field data capture unit. Students load a georeferenced aerial photograph (an aerial photograph mat has associate spatial location information for each point of the photograph) of the UAF campus on their Axim unit and then collect GPS locations around the campus. The combined GPS - Axim unit allows the students to visualise their location directly on the aerial photograph in the field. Later in the lab students download their data and include it in the larger database for use within a GIS.

Photogrammetry and Remote Sensing (two and a half weeks) - Earth system science relies heavily on spaceborne and airborne digital data. This unit introduces students briefly to the history, principles and techniques of photogrammetry, remote sensing and image interpretation. Aerial photos and satellite images showing diverse geographic features such as volcanoes, glaciers, coastal zones, impact craters, human settlements, forests, sea ice, are snared with the students and the students are encouraged to apply elements of photo-interpretation that they learned in the lectures, to visually interpret the images. This engaging exercise is then taken a step further and students are provided a time series of images to show processes such as coastal dynamics (erosion and deposition), glacial retreat, volcanic eruption, flooding, etc. Students work in teams of two to interpret the processes and their impacts. Students also carry out detail image interpretation of two images as a homework assignment.

In the third and the last lab period of this unit, students visit one of the satellite data receiving facilities at UAF to see how satellite data is received, processed, stored, and distributed to users in an operational environment.

Database Structuring and Management (one and a half weeks) - This short unit is used to present the concepts of spreadsheets and databases. Students receive guidance on effective use of a spreadsheet such as Microsoft Excel for data entry, storage, and export to other software packages. They also go through hands-on exercise on developing a small database such as a personal address database, a database of the music CD's they have at home, or a database of industries using geospatial data in Alaska.

Geographic Information System (three and a half weeks) - The GIS unit is the largest unit in the course. In this unit students learn about different data types, formats, compatibility issues, accuracy and quality. Lectures ground the students in the underlying principles of data storage, retrieval, integration, manipulation, and analysis. In the related labs students download geospatial data of different formats from the web sites of various data providers and researchers, convert the data formats to one they can use within their GIS, and then carry out spatial analysis using commercially available analysis tools. This unit also includes one to two guest lectures from GIS specialists on the use of GIS to address an Earth system science topic. In the last two years these lectures have included presentations on the hierarchical GIS for circum-arctic vegetation mapping, 3D GIS and terrain visualisation, and geodatabase designing for geological mapping. The unit also includes a visit to a local industry to show how GIS is being used in an industry setup. The industry visit also provides the students an opportunity to directly interact with a GIS professional and discuss future career options in the field.

Cartography and Geo-Information Visualisation (two weeks). The last unit of the course is dedicated to conveying the concepts of cartography and effective data visualisation. The students are given a photograph that depicts reality. They are then asked to represent reality as a hand-drawn map. Using student generated maps it is conveyed that map representations are imperfect and involve simplification, exaggeration, displacement, aggregation, omission, symbolisation and/or emphasis. In the associated lab, students also generate a map that is appropriate for computer screen display and a map that is designed for printing. One lecture in this unit is also devoted to providing tips on designing and delivering an effective presentation.

The independent student project is a major component of the course and constitutes 40 percent of the course grade. The importance, scope, and project expectations are discussed on the first day of the class, posted on the course website, and emphasized throughout the semester. Students work on their final projects in parallel with the structured weekly labs. A formal requirement of the course is that students submit a five to eight page project report, and make an oral presentation on the project results at the end of the semester. The final presentation topics and schedule are announced widely though the University's mailing lists and bulletin boards. The 15 minute time slot for each student includes 12 minutes for the actual presentation and 3 minutes for questions from the audience.

The advantages of project-based learning are well summarised at the starting point web site. The significance of project-based learning within the geoinformatics course stems from the great diversity in student projects which is commensurate with the diversity in student profile and interests. Some of the student projects completed in the last two offering of the course include designing a graphical user interface for entering and updating data into a personal geodatabase; mapping GPS measured deformation data for southeastern Alaska; spatial analysis of Humpback whale distribution and their behavioural patterns at Point Adolphus; linking Earth science research publications to geographic locations; and alternate route planning for the Trans Alaskan pipeline. It was our observation and conclusion that such diversity in independent projects stimulated curiosity and interaction among course participants who benefited by learning from their peers with different backgrounds and interests. In their interviews with an education evaluator the students concur that they "Recognised how similar techniques and tools can be used to investigate complex and completely different applications" and that the experience of going through tne entire process of selecting, planning, implementing and presenting an independent project was "the most demanding yet most rewarding part of the course" To assess the student performance in the course, the course includes a closed book quiz, a take home test with two essay type questions, home-work and lab assignments, and the independent final project report and presentation. The students are graded on an absolute 100 point scale where the quiz, test, homework/class participation, and final project account for 15, 20, 25 and 40 points, respectively. The grading is guided by a rubric, a matrix assigning points for each aspect to be assessed, that clearly describes the standards expected of students.

Other assessment instruments designed for this course included pre-course/post-course survey questionnaire; and a mid-semester/late-semester questionnaire and focus group. In the two offerings of this course, the pre-course questionnaires helped determine the starting point of students and provided a baseline to compare with the same questionnaires administered at course completion, to assess changes in attitude and ability. The mid-semester questionnaire and a mid-semester focus group discussion were administered in the absence of the instructor, by an external education evaluator as a formative evaluation instrument.

A second focus group session was held in conjunction with the post-course assessments. Assessment results show that in the two years of offering this course, all but one student showed an increase in understanding of how geoinformatics might be applied to their field of study. Students learned the basics of geoinformatics, and its applications. They were able to determine if additional courses in this area were desirable for their careers.

Though all assessment instruments play their role in helping to improve the course, as an instructor I found the combined outcome of the mid-semester questionnaire and focus group as the most important feedback that helped me to make timely formative changes in the course. How the course was extended beyond the University setting - The contents from the undergraduate geoinformatics course were adapted and extended to pre-service and in-service high school teachers in a special teacher's training course. Adaptations included condensing the semester-long course to an intense two-week course, shortening lab exercises, following a cookbook approach to lab assignments, and creating teams of two teachers to work together on the final project. Working in teams of two reduced anxiety among teachers and helped when they later tried to implement parts of the course in their classroom settings.

At UAF, the summer teachers training course was given in summer 2004 and summer 2005 and was taken by 14 teachers from different parts of the State of Alaska. Subsequently four teachers introduced components of the course in their high school science, mathematics and geography classes. The course is now being extended to school teachers in southeast Alaska with funding from the National Science Foundations' GeoEd Program.

Geoinformatics in the Educational Curricula

At UAF, the geoinformatics course was offered at the undergraduate level. However, the bigger question of when, where, and which components of geoinformatics should be integrated within the formal educational curricula remains unanswered. When should educators start laying the foundation of geoinformatics concepts?

Children begin forming spatial concepts early in their development. Comprehension of distance, direction, orientation, and relationship develops gradually as the child grows. Using visual exercises and three-dimensional models to reinforce these concepts in the kindergarten classroom lays the groundwork for using coordinates and graphing functions. Teaching these spatial concepts is already built into a child's early educational curricula in the United Sates.

Recent statistics on educational technology show that secondary school students in the United Sates have greater access to computers at school and home now than five years ago. Consequently, they are more computer savvy and better prepared to be introduced to components of geoinformatics in secondary school. Teachers who have taken the geoinformatics course at UAF have typically incorporated components of geoinformatics within their science, mathematics, and geography curricula, and have reported this to be a valuable supplement to the existing curricula. In post-secondary education, we at UAF experimented with introducing the geoinformatics course at the upper undergraduate level and found the students could successfully complete the course and benefit from it. The course at UAF was offered in the Department of Geology and Geophysics as UAF does not nave an established interdisciplinary undergraduate program. At other universities which have a setup similar to UAF, a geoinformatics course can be offered within any epartment, such as the departments of forestry, natural resources, geology, engineering, or computer sciences. If the university has sufficient resources to introduce a number of courses related to geoinformatics, such as courses in remote sensing, database management, cartography, programming, etc., geoinformatics can be offered as a specialisation within a disciplinary program, or as a standalone program in itself. At universities that have well established Earth system science programs or environmental programs, geoinformatics finds an ideal home within these interdisciplinary programs.

An upper undergraduate level introductory course in geoinformatics was established at UAF. The course was successfully completed by 16 students. The course contents were adapted and the re-tailored course was offered to 14 secondary school teachers in Alaska. The students who took the course came from different disciplines and they used geoinformatics to address completely different research questions and different Earth system science problems. Formative and summative course assessment tools were designed and used. Assessment results show that the course was successful in influencing student attitudes and abilities. Students developed specific skill sets as a result of this course that were useful in their research. The diversity in student profiles and student projects also showed that geoinformatics has the potential to be utilised as a capstone course that could be useful for students from many programs across the campus.

Does the Earth's shadow cause moon phases? What happens to water that evaporates in the water cycle does it dissociate into oxygen gas and hydrogen gas? These explanations for common Earth science phenomena are often offered by university non-majors, as well as the general population. Students may also exhibit conceptual difficulties in interpretation of topographic maps, aerial photos, geologic cross sections, and various two-dimensional diagrams showing three-dimensional or moving phenomena. Many misconceptions and broader conceptual difficulties have been reported in all four areas of the Earth sciences, although most are in the geosciences, astronomy, and meteorology. Although many conceptual problems are reported verbally as information-based misconceptions, such as the belief that seasons are caused by differences in Earth-sun distance at various times during the year, other conceptual difficulties are more broadly based. Examples are interpretation of geologic block diagrams, aerial and satellite photos, and topographic maps, as well as understanding of map projections, and of geologic time. In addition, the principles of the kinetic-molecular theory (KMT), which are basic to an understanding of meteorology, are also a source of many Earth science misconceptions. Other researchers have studied issues of scale related to Earth science, including models and astronomical distances, geologic distances, sizes of scientific objects, and time. Why are misconceptions important? Constructionist learning philosophy stresses that learners construct knowledge by assimilating new information with their personal previously held concepts. It is important, therefore, that students' previously held concepts are consistent with the ideas accepted by scientists, in order to establish a base on which to assimilate further concepts. The presence of erroneous previously-held concepts, or of broader conceptual difficulties, may impede understanding of Earth science concepts. Many Earth science misconceptions have been reported to be extremely difficult to change, even with thorough instruction and increasing age; this appears to be especially evident if the scientific explanation of Earth science phenomena is contradicted by intuitive knowledge derived from observations since early childhood.

Spatial ability has been defined as skill in "representing, transforming, generating, and recalling symbolic, nonlinguistic information". Spatial ability is a cognitive factor that has been linked to high performance in science and mathematics. A number of spatial abilities have been identified, although researchers have not reached consensus concerning names and descriptions of ability types, as well as categorisation of spatial ability tests in regard to factors measured. Some researchers fail to differentiate between ability factors when drawing conclusions, furthering the complexity of the literature.

In their meta-analysis, Linn and Petersen (1985) devised a classification of three types of spatial ability factors. They describe mental rotation as a "Gestalt-like" process involving "the rotation of a two- or three-dimensional figure rapidly and accurately". Although different strategies might be used to solve rotation test items, it is advantageous to be able to mentally rotate the figure as a whole rather than using more time-consuming analytic verbal clues, such as "if this side is turned here, then that side will be there." Linn and Petersen suggest the possibility that females, on average, may tend to use such a part-by-part analytic strategy, while males may tend to use a holistic analogue strategy to solve mental rotation problems. In addition, they suggest that females may be more cautious.

Spatial perception involves spatial relationships "with respect to the orientation of the subject's own body, in spite of distracting information". Spatial perception is partially characterised by the "possibility of relying on gravitational/kinesthetic cues". Linn and Petersen report that "The other feature of spatial perception tasks is a focus on disembedding or overcoming distracting cues".

Spatial visualisation is characterised by an "analytic combination of both visual and nonvisual strategies" that involves... "complicated, multistep manipulations of spatially presented information". These processes may involve the "processes required for spatial perception and mental rotations but are distinguished by the possibilities of multiple solution strategies". Linn and Petersen note that "These strategies are more characteristic of general ability than of spatial ability".

Of these three spatial factors, I hypothesized that mental rotation is most important in understanding Earth science concepts that are associated with common misconceptions and conceptual difficulties. In astronomy, humans are handicapped by their single vantage point from Earth of the moving bodies in outer space. Ability to mentally picture moving air masses from different perspectives is helpful in understanding meteorology, as well as ability to picture invisible moving molecules involved in phase changes. Viewing rock layers from varying perspectives is valuable in geology.

It is less clear how spatial perception is involved in Earth sciences. Possible connections of spatial perception to Earth science conceptual understanding include: a) ability to discriminate angular size, as in sun angles reaching Earth, compared to parallel "horizontal" solar radiation, in meteorology and astronomy, and b) disembedding patterns within in aerial or satellite photographs, such as drainage patterns.

Although a number of different types of tasks are present in various spatial visualisation tests (e.g., hole punch patterns on folded paper), the spatial visualisation test used in this research not only contains an element of differentiation of views from varying vantage points, but also involves two dimensional to three dimensional conversion. The latter is important in Earth science in interpretation of topographic maps and geologic block diagrams, as well as interpretation of any two-dimensional Earth science diagram that represents a three-dimensional phenomenon.

A large body of literature has described mean differences between the sexes, often favouring males, in some types of spatial ability. In general, the largest sex differences have been found in mental rotation, smaller differences have been found on spatial perception, and smaller or little difference has been found between the sexes on spatial visualisation. Apparently, most, if not all, tests of mental rotation tend to produce results with a male advantage; most researchers apparently do not consider this result due to gender-biased tests.

Interestingly, several other types of spatial ability are seldom mentioned in the science education or educational psychology literature. Of these, women tend to perform better with some (e.g., object location) and men with others (e.g., targeting). In addition, Signorella and Jamison (1978), Brosnan (1998), and Firth and Brosnan (2000) reported that psychological gender, the self-described possession of stereotypically masculine and feminine personality characteristics, is statistically related to spatial ability.

Mathewson (1999) and McCormack and Mason (2001) argue that spatial ability has been neglected in traditional education. Although students with weaknesses in verbal abilities are encouraged and given a great deal of practice, usually no specific help is given to students with weaknesses in spatial ability. Such students may eventually abandon the study of spatially-related disciplines, such as mathematics and science, when those topics become more difficult in middle and high school.

Public understanding of Earth systems and the ability of citizens to think critically about them is increasingly necessary as new technologies illuminate interrelationships among those systems, and as human influences on Earth systems intensify environmental concerns. University graduates who are non-science majors may eventually occupy positions in business, law, government, communications, and education, and therefore may greatly influence public policy on Earth resource issues. As policy makers, some may influence important decisions more than the scientists who advise them. It has been established, however, that non-science majors typically have more difficulty understanding various science concepts than do science majors, and also score lower on tests of spatial ability than do science majors. Non-science majors are less likely to be formal thinkers than are science majors, and tend to have both weaker study skills and more negative attitudes toward science than do science majors. Instructors of non-science majors, therefore, face many challenges in motivating and teaching this important group of future citizens. Only a few researchers have studied the relationship of specific types of spatial ability to specific types of Earth science concepts. I found no studies that attempted to relate Earth science misconceptions to spatial abilities. A number of researchers have reported, however, that spatial abilities can be improved.

Therefore, I thought that a need existed for additional knowledge concerning the relationship of specific spatial abilities to Earth science conceptual understanding by non-major university students. My hope is that, if such a relationship does exist, it may be possible to develop curricula for use throughout the K-16 educational system that may improve Earth science conceptual understanding through improved spatial understanding. I hypothesized that a relationship exists between many Earth science misconceptions and conceptual difficulties and spatial ability, and that mental rotation would be the type of rotational ability most highly related to scores on a test of Earth science conceptual understanding, and spatial perception least related. An understanding of scale also appeared to be part of a many conceptual problems (a situation that is likely not helped by models and diagrams that are not depicted to scale or explained as such), as well as difficulty in transforming two-dimensional diagrams to three-dimensional reality.

Subjects were non-science majors enrolled in six undergraduate courses in the College of Natural and Applied Sciences of a midwestern public university in the departments of geosciences, chemistry, physics, and biology. The courses were not designated for students majoring in science. I followed Institutional Review Board informed consent process with all subjects. Of 119 students who signed consent forms, I used data from the 97 students who were administered all five tests used in the study. Of these 97 subjects, 34 were male and 63 were female. Thirty were elementary/middle education majors, the most common single non-science major enrolled, 64 had declared other non-science majors, and 3 were undeclared. Of the non-elementary/middle education majors, most majors were related to business, finance, insurance, computer information systems, nursing and health professions, psychology, and various areas of education other than elementary/middle teaching. Subject ages ranged form 18 to 49, with a mean age of 24.42. Most subjects were upperclassmen, with a mean year of university study of 3.27, with 1.00 indicating freshmen. Earth science classes comprised only 10% of subjects' high school science coursework; 73% of subjects had taken no Earth science in high school. Earth science courses, however, comprised 34% of previous university science coursework. This may be due, in part, to a required Earth science course for elementary/middle majors previously taken by many of those majors.

Because I was unable to find any instrument with validity or reliability data for testing of Earth science misconceptions and conceptual difficulties, I developed and field tested the Earth Science Concepts (ESC) test. I describe the development, use, and results related to use of this instrument elsewhere. The 20 multiple-choice item questions and distractors are based on the Earth Science misconception and conceptual difficulties literature. Spatial ability factors were based on Linn and Petersen's (1985) classification. Other characteristics include Earth science disciplines involved, whether the item refers to a misconception or broader conceptual understanding, if a diagram is included in the test item, if the item refers to events normally observed throughout the observer's lifetime, and if scale or motion are thought to be involved.

All four areas of Earth science are represented, but to varying extents, primarily because oceanography is not highly represented in the misconception literature. Eight items include diagrams, and the others follow the verbal form used in the reporting of most misconceptions. Topics include the causes of seasons, moon phases, and tides, several phase changes as related to the KMT, geologic block diagrams, map projections, movements in the Earth/moon/sun system, relative distances within the universe, interpretation of aerial photos and topographic maps, relative distance to the magma that produces features such as volcanoes, and a three-dimensional drawing of weather fronts and air masses. I field tested the ESC fifteen times in non-majors science courses for over two years.

One class was a physics class and the rest were Earth science classes for preservice teachers. I administered a total of 310 pilot tests. Four versions of the ESC were field-tested, with 33, 32, 22, and finally 20 items. Scantron computer grading provided item analysis. I removed some items with very low or high item difficulty ratings, as well as some unselected distractors when new versions were constructed. I gradually removed some misconception items that appeared to be relatively knowledge-based and tended to replace them with items with diagrams and a broader conceptual base, as well as items that referred to phenomena that subjects witness from early childhood. I removed some items that were reported in the literature as children's misconceptions, but which were not troublesome with most college students in the pilot tests. Some distractors and items were added or changed as a result of interactions with students in my Earth science classes, including conversations and anonymous written answers to "How?" and "Why?" misconception-related questions I ask before beginning many new topics.

To ensure criterion validity, scores on the final version of the ESC that was field tested with Earth science classes were correlated with averaged Earth science content exam scores throughout the semester for the same individuals. The correlation coefficient was 0.60, which was moderately significant at a level of.01. To assure content validity, I asked content and science education expert faculty members to examine the ESC. Content faculty consulted were experts in geomorphology, geology, geographic information systems and cartography, meteorology, and physics. I made several revisions following their comments. I assessed the ESC several times for possible ambiguity, and removed one cartography item for this reason. I attached an open-ended question to field tests asking participants to describe any unclear or ambiguous questions. No subjects participating in the pilot tests did so. As I was the instructor of many of the students in the pilot tests, it is possible they did not want to appear unknowledgeable by saying a question was unclear. Furthermore, an "I don't know" option was available, which was designed to discourage guessing. Because no established Earth science conceptions instrument involving all four areas of Earth science similar to the ESC could be found, a construct validity coefficient could not be established.

Kuder/Richardson (K/R) and Spearman/Brown (S/B) split-half internal reliability statistics were computed in field tests of the final version of the ESC; Spearman/Brown was 0.74 and Kuder/Richardson was 0.63. Higher scores had been achieved with some previous pilot versions that were longer and that contained more knowledge-based misconception questions. Administration of the ESC to the 97 subjects in this chapter revealed reliability scores of S/B = 0.65 and K/R = 0.63. Higher scores could possibly be achieved if the test were longer and more homogeneous in content. A shorter test was developed due to the time restraints of administering a total of five tests to each subject. I administered timed paper and pencil tests that measured three types of spatial abilities, based as closely as possible on the three categories of spatial ability described by Linn and Petersen (1985).

All tests were cited in the science education literature. Two were commercially available, and I obtained the third from another researcher. Other criteria used to select spatial tests included measurement of spatial tasks that appeared related to avoidance of misconceptions or conceptual difficulties, and suitability for group administration. To test mental rotation, I selected the test called the Purdue Visualisation of Rotations Test (PVOR or VOR) on the test booklet and in the science education literature. The test is also called the Purdue Spatial Visualisation Test - Visualisation of Rotations and Purdue Spatial Visualisation Test: Rotations in engineering graphics literature, abbreviated as PSVT:R or as PSVT (Branoff).

The designation PSVT:R will be used here. In this test, subjects first determine the manner in which an initial drawing of a three-dimensional figure has been rotated to produce a second rotated drawing of the same figure. They then view a drawing of a second three-dimensional figure, and must then choose from five possible choices the drawing that shows the second figure rotated in the same manner as the first figure. Therefore, they must visualise the direction and extent of rotation of the sample figure before mentally rotating the second figure in a similar manner. Provo et al. (2002) relate that "the Kuder-Richardson and split-half reliabilities of the Purdue VOR test have been calculated in the range of 0.78 - 0.85 with samples of undergraduate chemistry students." Battista et al. (1982) found a K-R reliability of 0.80 when testing preservice elementary teachers. Branoff (1998) reports internal consistency coefficient results (KR-20) of 0.87, 0.89 and 0.92 from studies conducted by Guay (1980) on university students, machinists, and university students, respectively. Sorby and Baartmans (1996) reported a K-R coefficient of 0.82 for university students. Provo et al. (2002) reported that "the content validity of this instrument has also been shown to be good".

This test was chosen because it tested rotation of complex figures in three dimensions, which not only more closely simulates situations in Earth and space science, but is considered more difficult that two-dimensional rotation by several researchers. Of three three-dimensional rotation tests commonly mentioned in the science education literature, this test was more complex (a characteristic Linn and Petersen and Kimura have associated with performance) than one other possible test choice, Cube Comparisons and more readily available than the other choice, The Mental Rotations Test. I obtained test from Judy Provo, formerly of Purdue University.

Branoff (1998) developed a computer form of the PSVT:R that added x, y, and z coordinate axes to the three dimensions of each test item, with the intention of adding a contextual clue that might improve scores and response times. The addition of the axes had no significant effect on scores. Branoff states that "the addition of the axes eliminated gender difference on the PSVT". He later indicates, however, that only 17 females, who were divided further into the experimental and control groups, participated in the study. He also suggests that "one explanation for not finding gender differences for response times may reflect the type of students participating in the study. A majority of the students were enrolled in engineering programs. It may be that students in engineering, whether male or female, tend to take the same type of approach to solving mental rotation tasks". Therefore, students may not have been using the analytic approach, although the axes were designed to help persons who used the slower, part-by-part analytic approach.

Two other questions arise in regard to choosing whether to use Branoff's version of the test. First, is the altered test still measuring the ability of rapid mental rotation of a figure using the Gestalt-like process that Linn and Petersen described? They noted that "mental rotation items are used to measure the time required for solution" and "conditions of measurement encourage an analogue process". Secondly, does the addition of the letters x, y, and z, which add verbal cues, further change the basic nature of the test? Ironically, Piburn et al. (2002) developed a computer version of the Cube Comparisons to eliminate the preexisting lettered sides of the cubes, which they feared added verbal cues. I used the Group Embedded Figures Test (GEFT) to test spatial perception. In this test, subjects must locate a previously seen simple geometric figure within a larger complex geometric figure that has been organised to obscure or embed the simple figure within the complex figure.

Only straight lines that are horizontal, vertical, or diagonal compose either simple or complex figures. Because the subject must match the angles of the hidden figure with its duplicate within the complex figure, discrimination of angles or comparison to horizontal or vertical appears necessary to perform the task, as well as separation of figure from distracting ground. The test is the group-administered version of the original individually-administered parent test, the Embedded Figures Test (EFT). The manual for the GEFT relates that "the (EFT) is a perceptual test." Linn and Petersen (1985), however, do not list the EFT as a test of spatial perception, but as a test of spatial visualisation. They state, however, that "the other feature of spatial perception tasks is a focus on disembedding or overcoming distracting cues" (1482). They note a high correlation between the EFT and the Rod and Frame Test (RFT), an unavailable individually-administered earlier test of spatial perception and field dependence-independence which requires complicated equipment (a darkened room, and in one version, a tilted room). Field dependence-independence is described as the ability to discriminate relevant stimuli from distracting stimuli, the "field." Several other researchers disagree with Linn and Petersen's designation of the EFT. The two tests mentioned by Linn and Petersen as tests of spatial perception, the RFT and the water-level task are both individually-administered with physical objects, and neither is currently available. Therefore, the categorisation of the EFT and GEFT by their developers as a perceptual test was followed. Although some researchers devised paper and pencil water level tests, no reliability and validity data have been offered. Furthermore, it is unknown if any kinesthetic gravitational aspects of the physical-object tests would be as applicable to paper and pencil tests.

Witkin el at. (2002) reported that because the GEFT is a timed test, an appropriate method of estimating reliability is the correlation between parallel-forms with identical time limits. Correlations between the nine-item first section scores and the nine-item second section scores were computed and corrected by the Spearman-Brown prophecy formula, producing a reliability estimate of 0.82 for both males and females. Reliabilities for three groups of college age students completing the parent EFT were 0.82 for males and 0.79 for females in the first group, 0.90 for males and 0.82 for females in the second group, and 0.85 for a third male-only group. Regarding validity, Witkin et al. stated that "the most direct criterion measure is the 'parent' form of the test, the EFT." Correlations of the GEFT with the EFT are -0.82 for males and -0.63 for females. The validity coefficients "should be negative because the tests are scored in reverse fashion". Additionally, the GEFT and EFT have exhibited criterion validity by correlations with the RFT. Researchers who have reported use of the GEFT include Piburn et al, 2002; Kwon and Lawson (2000), Brosnan (1998), Firth and Brosnan (2000), and Signorella and Jamison (1978).

Discrimination of angles is likely more important in understanding spatially-related concepts on the ESC and in Earth science than is the specific ability to discriminate vertical and horizontal lines from lines at other angles. In comparison to mental rotation and two- to three-dimensional transformations, relatively few concepts and commonly held misconceptions in Earth science appear to directly involve disembedding skill, although possible associations include map and aerial and satellite photo interpretation. I tested spatial visualisation using the Differential Aptitude Test: Space Relations (DAT). I selected this test due to the required task of two-dimensional to three-dimensional transformation.

The test asks the subject to visualise a two-dimensional pattern as it would look after being folded into a three-dimensional figure, and then to select the correct drawing of the three-dimensional figure from choices that are viewed from various perspectives. This type of test is often termed a surface development test. Other types of spatial visualisation tests, such as paper folding and punching tests, do not seem as applicable to Earth science concept understanding. Bennett et al. (1991) report a K/R internal consistency reliability of 0.94. They also report a part-whole correlation of the formerly used DAT Form V with the currently available shortened version, that was used in this research, of 0.96. Criterion validity was obtained by examination of correlation coefficients between DAT test scores and criterion scores on job training tests. These included 0.59 for draftsman applicants, 0.49 for freshman engineering students, 0.55 for dental hygiene students, and 0.52 for electricians. Construct validity was shown by a correlation coefficient of 0.68 with the spatial relations subtest of another multiple aptitude battery, the General Aptitude Test Battery.

Earth science involves a number of concepts that require mental visualisation of three-dimensional phenomena, which may be viewed from various vantage points, using two-dimensional representations; such concepts appear to require skills similar to those used to answer DAT items. These include understanding of geological block diagrams, air and satellite photos, map projections, and contour maps. Researchers who have reported use of the DAT include Feingold (1988), Provo-Klimek and Cash (2001), Huit and Brous (1986), and Fennama and Sherman (1977).

Subjects also completed a questionnaire that queried them about demographics and past science coursework. I administered the short form Bern Inventory (BI) to determine the psychological gender category and BI-t score of each subject. I administered this test due to questions raised by past researchers.

ESC Scores — Many Earth science misconceptions and conceptual difficulties were revealed in this population of university non-science majors. The mean score for these non-majors on the 20 item ESC was 6.82, or 34%, with a minimum of 1 and maximum of 16 (SD = 3.03). Difficulty ratings, which indicated the percent of students selecting non-scientific explanations, were determined for each item. Ratings ranged from 96.42 to 33.04.

Spatial Ability and ESC Scores — Significant positive correlations at a moderate level between scores on the test of Earth Science concept understanding and scores on each of the three spatial ability tests. For mental rotation, (r(97) = 0.52, p = < 0.01. Subjects who scored higher on the PSVT:R tended to score higher on the ESC. For spatial perception, (r(97) = 0.34, p = < 0.01. Subjects who scored higher on the GEFT tended to score higher on the ESC. For spatial visualisation, (r(97) = 0.43, p = < 0.01. Subjects who scored higher on the DAT tended to score higher on the ESC. Low to moderate correlations were found among the three spatial ability tests.

Stepwise regression was conducted first using all three spatial test scores as independent variables, and then using the three spatial test scores and number of completed university Earth science courses as independent variables. In both procedures, models were produced that indicated that PSVT:R score is the independent variable that is the best predictor of ESC score, of the variables used in the regression. In the first model, with the three spatial tests as independent variables, an R value of.52 indicated as estimate of a 52% relationship between ESC score and the best linear combination of independent variables. The R2 of .27 indicated that 27% of the total variation in ESC scores can be accounted for by the three types of spatial ability. In the second model, with the three spatial tests and number of past university Earth science courses as independent variables, an R value of .56 indicated as estimate of a 56% relationship between ESC score and the best linear combination of independent variables. The R2 of .31 indicated that 31% of the total variation in ESC scores can be accounted for by the three types of spatial ability and number of past university Earth science courses.

Demographic Analysis — it shows correlations among all tests and most demographic variables. I found a weak correlation between number of previous university Earth science courses and ESC scores. Statistical analysis revealed that neither sex, age, university grade level, nor major (preservice elementary/middle majors in comparison to all other majors) was significantly related to scores on the ESC or any of the spatial ability tests. Independent samples t-tests determined if a significant difference existed between female and male test scores on tests and demographic variables. Results revealed that although males outperformed females on the ESC, the PSVT:R, and the GEFT, the differences were not significant. Females and males performed nearly identically on the DAT. Females had taken significantly more science classes at the university level, were significantly older, significantly more likely to be elementary/middle education majors, and had significantly higher Bern numbers and T-scores, indicating greater femininity.

Pearson correlations of Bern Inventory T-scores revealed no significant relationships between psychological gender with any of the three spatial ability scores or ESC scores. Further t-tests comparing each possible combination of the four psychological gender categories in regard to all other demographic characteristics and test scores revealed no significant relationships between psychological gender and test scores.

A significantly positive correlation was found between number of completed Earth science courses at the university level, age, university grade level, and elementary/middle education major, as well as ESC scores. No significant relationships were found between number of completed high school or university science courses or number of high school earth science courses with any test scores.

The results of this chapter provide preliminary data that suggest a possible relationship of selected spatial test scores and Earth science misconceptions and conceptual understanding. They are also consistent with the proposed idea that mental rotation is a type of spatial ability that is associated with a number of Earth science misconceptions and conceptual difficulties. Of the variables tested, regression analysis indicates that between one fourth and one third of the total variation in ESC scores can be accounted for by spatial ability, an ability that has not been systematically fostered in traditional education.

Low to moderate correlations among scores on the three spatial ability tests is consistent with the notion that all are measures of spatial, rather than verbal ability. They were not highly correlated, as they might be if they were measuring the same spatial abilities.

PSVT:R score, or mental rotation, is the variable studied that best predicted ESC performance. This result is consistent with the aspects of movement involved in many Earth science concepts on the ESC, and the necessity to visualise the position of objects from varying vantage points. Some of the most persistent misconceptions exhibit these characteristics, including the mechanism of moon phases, the mechanism of seasonal change, and the invisible movement of molecules during phase changes. The PSVT:R requires three-dimensional thinking, which is also required in understanding these concepts

Many subjects completed the PSVT:R before the suggested 20 minute time limit had expired. This may have somewhat diminished the capacity of the test to measure the ability of subjects to rapidly and accurately rotate objects mentally in a Gestalt-like manner, rather than in an analytic manner. Linn and Petersen suggested that females may, on average, tend to use analytical strategies rather than a holistic Gestalt-like ability, which is more common with males. This time limit may have allowed more time for persons who tend to use analytic strategies to do well on the test. This chapter's result of a nonsignificant male advantage on the PSVT:R is inconsistent with most findings that males exhibit a significant advantage on all tests of mental rotation. It is unknown if this time limit was related to this result. I have seen no other references to the time limit in the literature.

DAT results are consistent with the idea that conversion of two-dimensional images to three-dimensional reality in the Earth sciences is a component of several misconceptions and conceptual difficulties, as well as the viewing of phenomena from different vantage points. The highest correlation between spatial ability tests occurred between the PSVTiR and DAT. This is not surprising, as both have components of viewing three-dimensional objects from differing vantage points, although other aspects of the tests are not similar.

A contrasting pattern of DAT results compared to PSVT:R and GEFT results is evident in several respects. Males outperformed females on the other two tests, although the advantage was not significant, but male and female scores on the DAT were nearly equal. This result is consistent with the observation that spatial visualisation tests are characterised by use of analytic strategies and often show little, if any, difference in male and female performance.

The possible relationship of spatial visualisation to verbal factors and general ability indicated by Linn and Petersen (1985) may also be partially responsible for the disparity between mean percentage scores on the three spatial tests. Mean scores on the PSVT:R and GEFT were 52% and 48%, respectively, but the mean score on the DAT was 67%. This represents 15% and 19% higher mean scores on the DAT than on the PSVT:R and GEFT, respectively.

A contrasting pattern of DAT results compared to PSVT:R and GEFT results is also evident in regard to past science coursework. DAT scores were significantly related to the number of Earth science classes completed at the university level, but the other two tests' scores were not. Also, elementary/mid die majors, who had taken significantly more university Earth science and science courses than all other majors, scored quite similarly to all other non-science majors on the ESC, PSVT:R and GEFT, but scored higher on the DAT than all other non-science majors, although the differences were not significant. These differences between results on the PSVT:R and GEFT in comparison with the DAT are consistent with Linn and Petersen's (1985) description of spatial visualisation in comparison to mental rotation and spatial perception. Linn and Petersen observed that spatial visualisation depends on meta-strategy for selecting processes rather than an individual's relative proficiency for using any particular process. As mentioned, they considered visualisation strategies to be more typical of general ability and more closely related to verbal factors than are mental rotation and spatial perception strategies. University coursework may help develop some of these more general, and more verbal, strategies.

Most preservice elementary/middle teachers in the study had completed the required Earth Science for Teachers course, a fact that contributed to the significant correlation between that major and number of completed university earth science and science courses. Elementary/middle majors in this chapter also showed significantly higher university grade level, which is perhaps indicative of the tendency of many preservice teachers to delay taking physics and Earth science courses until later semesters in their university careers, as well as the biology prerequisite. Because the great majority of elementary/middle majors were female, it is possible that some relationships exist between the additional Earth science and university training received by this subgroup of subjects and female performance on some of the tests. To further assess gender and major differences, further investigation may be indicated of preservice teachers at both earlier and later years in their university careers, as well as non-major populations that do not include preservice elementary /middle teachers.

Only a weak correlation, the lowest correlation between any two of the spatial ability tests, was found between the DAT and GEFT. This finding is consistent with the use of the GEFT as a test of spatial perception, rather than a test of spatial visualisation, as suggested by Linn and Petersen (1985). It is also consistent with the observations of several researchers that disembedding may not be classified as spatial visualisation. In addition, the GEFT requires only two-dimensional thinking, while both the DAT and PSVT:R require three-dimensional thinking. Linn and Petersen's description of spatial visualisation includes "multistep manipulations"; Eliot (1980) considers embedded figures tasks to be matching, rather than manipulation tasks, as the subject must match angles and line lengths.

Students with ability to discriminate angles and discrepancy from vertical or horizontal may find it easier to understand Earth science concepts that involve angles, such as sun angle during different seasons and the relative angles of the ecliptic and Earth's equatorial plane, as well as bedding planes and strike and dip. Disembedding may be related to ability to discern individual features on aerial or satellite photographs. Much work remains to be done, including further delineation of spatial ability factors and tests. Not only is there a great deal of overlap and ambiguity among classification methods proposed by various authors, but other researchers have identified possible additional spatial factors. Kali and Orion (1996), for example, have described a visual penetrative ability, VPA, which they have associated with successful interpretation of geologic block diagrams, and which they consider a subset of spatial visualisation.

Spatial visualisation itself is a somewhat nebulous category. Linn and Petersen (1985) do not delineate spatial visualisation by the same criteria as spatial perception and mental rotation, which apparently involve specific abilities. They describe spatial visualisation, instead, as often using the same abilities attributed to spatial perception and mental rotation. They delineate spatial visualisation by choice of multiple strategies rather than by ability type.

Much of the development of spatial ability tests used in education and a large body of research occurred in the 1970s and 1980s, and subsequently a number of cited tests are not now easily available to researchers. Commercially available tests are primarily used for employee testing. Although research on spatial ability has continued in the following decades, little additional development of tests has apparently occurred, other than the production of group-administered forms of several tests. A need exists for both clarification of factors measured by specific tests and development of tests.

Much additional research is implicated by this preliminary study, including a similar study using many more subjects. I am analysing any relationships between individual ESC items and scores on individual spatial ability tests, and intend to continue this chapter as I collect more data. I am also currently planning and piloting a number of spatially-related activities for use in my Earth Science for Teachers course that I will implement next year, as part of a university grant, and will employ the three spatial tests and the ESC as pre- and post-tests. I also plan to test preservice teachers at an earlier stage in their university careers.

Understanding of scale in both distance and time appear to be related to many Earth science conceptual problems. Several researcners have begun work concerning the cognitive basis of student understanding in this field (e.g Dodick and Orion, 2002, with geologic time). Additional work will hopefully be enlightening. Although many factors are certainly associated with understanding of Earth science concepts, the contribution of spatial ability that is suggested by these results should not be overlooked, nor should studies that indicate that spatial ability can be enhanced. A word of caution may be in order, however. Although research indicates that spatial ability can be improved, those results, as well as many others concerning spatial ability, may require further research and clarification. Some studies do not differentiate between types of spatial ability (Linn and Petersen, 1985). A few intervention studies that do differentiate between spatial ability factors found improvement in some spatial or Earth science abilities, but no improvement in mental rotation. Interestingly, mental rotation is the factor I found that best predicted ESC scores. Many questions remain, such as the possible influence of age at which interventions begin. A great deal of future work in many related areas is indicated, and could prove exciting. Therefore, this chapter's results suggest that an opportunity may exist to improve Earth science conceptual understanding by development of curricula and interventions that focus on spatial aspects of concepts.

Perhaps the neglect of educators to foster spatial ability is related to its history of association with practical, rather than academic, skills. Perhaps educators have assumed that nothing can be done to improve spatial ability. The known association of success in science with spatial ability underscores the importance of additional research and of attempting a more deliberate and informed focuses on spatial aspects of science concepts by K-16 teachers.

8

Integrating Cathodoluminescence

Cathodoluminescence (CL) has traditionally been used in carbonate petrology studies but this technique is rapidly expanding into clastic sediment provenance studies. CL can also be applied to high temperature geologic materials such as igneous and metamorphic rocks, ore deposits, gems, and geochronology. On the more applied side of science, the Federal Bureau of Investigation relies significantly on CL image analysis not only for provenance of traditional samples such as soils, concrete, and sand but also nontraditional samples such as paint, explosives, and duct tape. The wide range of CL applications in the geosciences creates the opportunity to incorporate the technique in a variety of courses within the undergraduate Geology curriculum.

Across scientific disciplines, colleges and universities are increasing the integration of instrumentation in undergraduate education. The use of instrumentation in undergraduate education increases awareness to the real-world impacts of science. Instrumentation also provides a visual approach to comprehending data that can increase understanding beyond traditional methods of data presentation. The use of instrumentation early in an education allows students to become familiar with techniques that will be valuable not only throughout their undergraduate career but also in graduate school or industry positions. Knowledge of modern analytical techniques and proficiency in their uses are important goals for undergraduate programs. CL has been only nominally mentioned in the Geoscience education literature despite its widespread use in research. Kopp (1981) presented an introduction to the applications or CL in geological research, but not a geology curriculum. Amthor (1993) provided examples of how CL imagery is used for carbonate petrography, in conjunction with backscattered images, but included no discussion on how to incorporate this technique beyond a single application. The purpose of this chapter is to provide details on how CL has been incorporated into two courses at Illinois State University and now this can serve as a springboard for use in other courses in our Geology program. All the exercises and assignments described here are for the Luminoscope ELM-4 with a Diagnostic Instruments SPOT Insight Firewire digital camera at Illinois State University. The benefits of the ELM-4 are that students can see true colour CL emissions and the equipment is nearly an order of magnitude less expensive to acquire than a scanning electron microscope (SEM) (approximately $25,000 for the ELM-4 and camera if an appropriate microscope is already in place).

Cathodoluminescence in Mineralogy

Theory of CL: Much like X-ray diffraction, X-ray fluorescence and optics, the theory behind cathodoluminescence is introduced in Mineralogy. A working knowledge about the science behind the process is important for students to understand. The CL material is covered after lectures about colour in minerals, introduction to the petrographic microscope, some mineral groups, and crystal chemistry. At ISU, Mineralogy students have had at least two previous geology courses so have familiarity with common rocks and minerals and common geologic processes. Students, therefore, are familiar with viewing minerals in plane polarised and crossed polarised light. The students are also familiar with interpreting colour spectra from the lecture about colour in minerals. This background has been sufficient for students to begin the CL work described below.In class, I begin the CL lecture with images of minerals viewed with different techniques. The students can then become aware that grains of zircons in a binocular microscope can look very different than a grain mount in plane polarised light, cross polarised light, and CL. All images may give worthy information, but depending on the questions being asked you may need to look at a geologic sample using a number of different methods.

After viewing the images mentioned above, I then define luminescence. I give examples of different types of luminescence, phosphorescence and fluorescence, and define cathodoluminescence as a type of fluorescence requiring an electron beam as an energy source. I describe that the electron beam strikes the surface of the sample and that some electrons are absorbed by the sample while some electrons interact with the sample. The CL emissions we observe, either with directly through a microscope or imaged with a camera, are those resulting from electron interaction in the sample that occur within the visible part of the electromagnetic spectrum. Topics similar to those previously covered in class on the source of colour in minerals, such as the idea of electrons being excited from the filled band up to the conduction band and subsequently falling back down to the filled band, are mentioned. The concept of activator elements and quencher elements are mentioned. The class has already covered electron configuration and electron shells so the application of these concepts is not a big step for the students. Students can start to understand that some elements are more likely luminescent and therefore two grains of the same mineral may show luminescence differently because of minor differences in chemistry. Some examples include Fe and Mn content controlling the intensity of CL in calcite. Examples of typical quencher elements and the associated colours of activator elements in minerals can be discussed as well. Some examples include Sm^{3+} and Dy^{3+} in anhydrite producing a tan CL emission and Mn in anhydrite producing a green emission. I urge the students to pursue mineral specific activators and quenchers in later projects.

While the Luminoscope at ISU does not permit spectral analysis of CL emissions, I do show CL spectra in order to place context for how geologists identify the activator elements in particular minerals. I discuss the peaks in various spectra that occur with wavelengths for different elements and mention that these emission spectra are used to identify the activator elements in a particular mineral or identify mole fraction of endmember compositions in solid solution (e.g. plagioclase). Again, the class has already spent a full lecture on colour in minerals so the spectra are not a new concept. To tie in other crystal chemistry concepts from mineralogy, mention can also be made about how the same element may behave very differently in different minerals depending on the bonding type and mineral structure. Ferrous iron is a typical CL quencher element but Fe^{2+} has been shown to be an activator element in experiments with feldspars.

Applications of CL: Having presented the science behind CL, I then emphasize why CL is such an important tool for many geologic applications. The colour and pattern of a particular mineral's CL emission can provide information about the temperature of mineral formation (e.g. authigenic vs. igneous quartz), diagenesis or alteration history (e.g. quartz, carbonates), zonation of chemical composition (e.g. zircon, garnet, apatite, fluorite, pyroxene), and possibly evidence of fluid-rock interaction. In class, I provide three different scenarios and I expect the small groups of students to be able to interpret their scenario based on the introductory lecture material discussed above. These exercises are discussed in the second half of an hour-long lecture period. In the past I have had each group of students report the interpretations of their scenario to the rest of the class in the final minutes of class. This can seem rushed so in the future I will have the students present their findings at the beginning of the next class period. The extended time will give them a chance to formulate a more complete story and achieve a better understanding of the scenarios. Before the students informally present the results to the rest of the class, I confirm the completeness of the interpretation but I do not grade the results.

Fluid-Rock Interaction: This first scenario emphasizes fluid rock interaction in rhyolitic rocks of a volcanic-hosted massive sulfide deposit. The students are given the following information along with the geologic map and CL images from King et al. (2000):

You are a geologist working at Sturgeon Lake, Ontario, a lead/ zinc ore deposit. The ore is hosted in a sequence of volcanic rocks, mostly rhyolites. The stratigraphy of the volcanic units has been miraculously preserved in the original order, despite the fact that the rocks are all Archean in age. The entire package of volcanics has been tilted on its side so the older units are to the south and they young northward. You are particularly interested in quartz phenocrysts within two units, the Mattabi and the Tailings Lake. Remember that the Mattabi unit is host to one of the major deposits and the Tailings Lake unit, located just below the Mattabi unit, is unmineralised. Describe what you see in the two pairs of images and then come up with a reasonable geologic story to correlate the evidence in the images and the location of the ore deposit.

The students develop a basic interpretation that the ore deposit involved fluids and that magmatic quartz phenocrysts recrystallised with later stage quartz in cracks and veins. The pattern of recrystallisation is only visible in CL because of different concentrations of activator trace elements in the two stages of quartz formation. The barren unit has not witnessed the same extent of fluid interaction, if any, so the quartz phenocrysts do not display fracture filling.

Authigenic and Detrital Quartz: To demonstrate the different CL colours based on temperature of formation, a second scenario involving the St. Peter sandstone is presented using Plates 4a and 4b from Marshall (1988) and the following information:

You are a geologist working with the St. Peter sandstone. The sandstone is Ordovician in age and located in Minnesota, Missouri, Wisconsin, Indiana, Illinois, Kentucky, and Arkansas (so much of the

Midwest). The sandstone is made of fine- to medium-grained well-rounded and well-sorted frosted grains of quartz that are weakly cemented. This sandstone can be a major aquifer for towns and cities.

In some locations, interesting textures are visible in both crossed polarised light and in CL. This sample, from Arkansas, consists entirely of quartz so you don't need to worry about identifying any other minerals. Your job is to describe what you see in the images and them come up with a likely story to explain the evidence. Try to create a reasonable geologic story. The students Recognise that there are overgrowths on the quartz grains and the CL is useful for distinguishing temperature of formation. In the lecture material before these exercises, I did show the spectra of quartz formed at low and high temperature and mentioned the typical colours associated with them. In the St. Peter Sandstone there are authigenic rims with brown luminescence precipitated on detrital quartz grains with a blue luminescence. The students will refer back to the lecture notes and be able to Recognise the centre of the quartz grains formed at high temperatures (the detrital component) and the rims show the typical colour of quartz formed at low temperatures (the authigenic component).

Geochronology: CL imaging is a very common technique applied to zircon U-Pb dating to identify detrital grains, metamorphic overgrowths, magmatic oscillatory zoning, and U contents. Inherited zircon cores are very easily observable in CL so the following scenario is presented to the students for interpretation along with the geologic map and images from King and Valley (2001):

You are a geologist working in the Idaho Batholith, mostly granite with an intrusion age ranging from 100 to about 50 million years. The batholith is surrounded for the most part by Proterozoic Belt Supergroup sediments that are between 1070 and i860 million years old. In the lab you are looking at zircons separated from two samples from the batholith. The zircon ages from these samples suggest ages between 1860 and 50 million years old.

You see some interesting textures in zircons in CL that aren't apparent in crossed polarised light. Everything in here is zircon; there are no other minerals you need to worry about identifying. Your job is described what you see in the images and then come up with a likely story to explain the evidence in the images. Try to create a reasonable geologic story.

Based on this simplified scenario, the students determine that the granite ascended through and some samples incorporated the Belt

Supergroup. Some zircons consist of assimilated detrital cores with magmatic rims while others represent a single growth episode.

From these three geologic applications, the students at least get an initial taste for the breadth of the uses of CL. They not only realise that viewing minerals in different ways (plane light, cross polarisation, CL) can display different information, but that with only a few images a much more intricate geologic history can be determined.

Instrumentation: To introduce the instrumentation, the students were shown pictures and schematic drawings of the ELM-4. As with most instrumentation, however, they don't truly understand the equipment until they have hands-on experience. Because the Luminoscope is often the first major piece of instrumentation that the students experience, I introduce the ELM-4 in stages to avoid overwhelming them. The first step emphasizes the image capture part of the process. During the same lab in which I cover a basic introduction to the digital image capture system on the traditional pétrographie microscopes, I walk the students through the Luminoscope imaging system. I also use this supervised opportunity to walk them through how to focus the microscope, cross the polars, switch the light source between the eye pieces and the camera, and the location of the few other parts of the microscope they will need to use since the ELM-4 sits on a different microscope than the ones they use in class. To ease the students into the image capture software, we start with simply acquiring images in plane polarised or cross polarised light, not the low light situation required for CL imaging. I have them launch the software, learn how to view a live image, change the contrast/ brightness/quality settings, and capture an image. In about twenty minutes the students have acquired a moderate comfort level with the software. The students turn in a PowerPoint file with lebelled images from the Luminoscope in addition to their first images captured from the polarising microscope.

The second step of the introduction to the Luminoscope is a homework assignment following the step-by-step procedures established for start up and shut down in order to become familiar with the vacuum pump, sample chamber, and software that controls the vacuum and beam current. This assignment is completed out of class and working with a partner or small group is encouraged for moral support. The start up/shut down takes less than thirty minutes for the students to get oriented with the sample chamber, vacuum pump, and software interface. Use of the logbook is also emphasized at this stage so I can keep track of Luminoscope use and problems as the semester progresses. The homework assignment is to record their work in the logbook and turn in a brief write up with their comments about the assignment and how clear they felt the directions were. This homework allows me to monitor any major student fears or problems working with the instrumentation and tinker with the instructions if I am missing a step or description that is clear to me but confusing to them. Independent, unsupervised student access is possible by providing a key to a locked lab during the portion of the semester the students are using the Luminoscope. The initial lab encounter with the ELM-4 is usually enough to instill significant respect of instrumentation.

The only remaining portion of the Luminoscope instrumentation that the students haven't experienced now is turning on the electron beam and acquiring an image of the mineral in low light CL conditions. Pumping down the sample chamber to achieve the appropriate vacuum level for the electron beam takes approximately 30 minutes and this pump down time is skipped initially to streamline the process. Integrating the sample chamber pump down, electron beam use, and imaging software occurs in the Pet Mineral Project described below. After the instrumentation introduction described above, the students are encouraged to complete the project on their own schedule with their partner. In case problems arise I encourage the students to work when I am around for help, but more often than not the CL image can be captured without further assistance. The electron beam work is the most straight forward so if the students are comfortable getting the sample in the vacuum chamber and using the image acquisition software they are able to be independent in the lab.

Pet Mineral Project: The Luminoscope is nicely included as part of a semester long project assigned to students in Mineralogy. On the first day of class each student receives an unknown mineral sample. Unbeknownst to them, there is a partner in the class with the same mineral, but often a very different looking hand sample. During the semester, students must determine the mineral's physical properties, create an image of its structure using a crystal structure software program, describe the symmetry, capture images in plane light and cross polars, and finally capture an image of the mineral in CL. After determining the physical properties of the sample on the first day, leading to mineral identification, students realise they have a partner in class and the rest of the project is completed together. Initially all I pass out are hand samples of each mineral but I have thin sections, both polished and covered, available for the students when needed.

The final project is submitted in the form of a PowerPoint file containing all the images and data acquired throughout the semester.

Clearly minerals that produce bright CL need to be specially selected to make the CL image acquisition as easy and interesting as possible. Minerals that have worked well so far include calcite, dolomite, sillimanite, wollastonite, microcline, clinopyroxene, and apatite. Other possibilities include andalusite, aragonite, favourite, and sodalite. Jadeite and corundum also display clear CL. This mineral list is by no means meant to be exhaustive but to demonstrate that common and typical rock-forming minerals are luminescent. Students don't necessarily need to focus on the more obscure luminescent minerals (e.g. amblygonite, benitoite, cuprite, or smithosonite) because it is possible to incorporate minerals commonly encountered in the mineralogy classroom into the CL work. Certainly for departments with a SEM equipped with an energy dispersive spectrometre, this project can easily be carried one step further with chemical analysis if any zoning is observed in minerals.

An exercise demonstrating the importance of CL for geochronology is easily introduced into a geochemistry course and is not a large undertaking if the students have already taken Mineralogy. The students are given two zircon grain mounts, both from the Idaho batholith. One sample displays magmatic growth zoning and the other contains inherited cores. The assignment asks the students to capture images of representative grains from each sample. The acquired images should represent the presence of and lack of zircon cores in the respective samples. Based on these images, the students are then asked to discuss what issues there may be for dating each of the samples. Students must address questions such as: How meaningful would an age be if analysis were done on a collection of grains or a single grain? Is an ion microprobe (Sensitive High Resolution Ion Microprobe a.k.a SHRIMP) necessary to obtain meaningful ages or can more conventional techniques be used? If the students have already taken Mineralogy and gone through the step-by-step introduction to the instrumentation, this homework assignment can be handed out and be completed within a reasonable time period.

A second part of the assignment involves the image of the zircon from the Idaho Batholith mentioned earlier. This is used as an example in order to give the students an idea of the challenges geochronologists can face with zircons that experienced multiple growth events. The students are asked "In the zircon seen in CL, also from Idaho, the core fills roughly 20% of the volume of the grain. Considering that the age of the core is approximately 1,800 Ma and the rim is approximately 90 Ma, what would the apparent age of the entire grain be? If you analysed this grain as a single, entire grain, would the age be significant or troubling?" This combined exercise of using the instrumentation and interpreting CL images reinforces the technical aspects of using the Luminoscope and the importance of applying CL imaging to this particular type of geologic problem.

Only minimal adaptations to the exercises described above would be needed if students acquire CL images on a Scanning Electron Microscope. The main change would need to occur with the instrumentation introduction exercises. This portion could still be completed in two parts with perhaps part one as an introduction to insertion of samples in the chamber and pumping down to vacuum and a second exercise to experience the imaging part of the software interface. It is important to note that the incorporation of CL work in an undergraduate Mineralogy course does not require much additional room in the syllabus. I have one dedicated lecture to cover the theory behind CL in addition to the scenarios used to introduce the various applications of CL in Geology. I added the homework assignment for the startup of the instrumentation. The image capture introduction was added on to a lab that was already relatively short so no additional lab time was needed for this portion. The Pet Mineral project was already in place in the syllabus so the only addition there would be the capturing of the mineral in CL. After all the other CL work, the students only need another hour at most to capture a CL image of their Pet Mineral. The theories and concepts behind CL easily augment traditional techniques taught during Mineralogy and can be a natural extension of any current syllabus. The Geochemistry course was new so no modifications of an existing syllabus were needed.

Students Recognise the importance of learning instrumentation in their geologic studies, even before being exposed to most of the various instruments and applications. When asked at the beginning of Mineralogy, the first intensive course in the program for most majors, to rank on a like-ard scale of 1-5 (5 being very important) the importance of learning to use instrumentation as a component of a geological education, the average result was 4.9. When asked the same question at the end of the semester, the average response from the same group of students was 4.7. The average response to "rank how important it is to use instrumentation for characterisation or investigation of geologic samples" was 4.6 out of 5 (5 again being most important) at the beginning of the semester and 4.8 at the end. When asked how comfortable the student is using instrumentation in the lab setting, the average student response was 3.7 at the start of the semester and 4.2 at the end. The question about the importance of using the newest technology to acquire skills not traditionally covered in courses brought an average response of 4.2 at the beginning and 4.5 at the end. At the end of the semester, only 15% of the respondents said that they didn't feel they had a basic understanding of how CL works. When asked if the student would be interested in working more with CL for research or coursework, 85% of the responses were positive.

Only one question saw a significant increase or decrease over the course of the semester. The comfort level that students have with using instrumentation increased from 3.7 to 4.2. This increase shows the success of the step-wise introduction to the instrumentation rather than overwhelming the students with a quick, full immersion approach. Students will be more inclined to participate in research involving instrumentation if they are not afraid of it. I feel the introduction of CL in this way was successful based on the increase in the average response.

Quantitative assessment on the incorporation of CL into the undergraduate curriculum is difficult, but qualitative feedback from students and additional uses of CL for independent studies can be considered. A post-survey was administered at the end of Mineralogy to assess various aspects of the class. Three of the questions were: "How did you like the Luminoscope CL exercise? Was it worthwhile to do? Would you be interested in doing more work with CL?" Vague negative responses included "no, no, and no", "I did not understand it or its use" and "it was disappointing". More constructive negative comments included "Yes, but it was hard not doing it with you", "It was ok, I was afraid of messing up" and "I liked learning about it but it was frustrating to use". This frustration and fear is not a surprise from students that have encountered very little instrumentation before Mineralogy. Use of the Luminoscope requires the use of two software programs on different computer platforms and slightly finicky hardware that all require a bit of intuition to anticipate quirks and problems. The remaining nine responses from the survey contained only positive feedback about the experience and included surprising comments such as "It wasn't worthwhile as we didn't do enough of it" or "it was a good idea, but more should have been done with it". Similar attitudes of wanting more time with the instrumentation has been documented by others introducing instrumentation in undergraduate courses.

Two students at ISU have completed independent studies using CL. One project created the user-friendly instructions for startup, shutdown, and image capture. This was completed before CL was introduced into the Mineralogy syllabus. The other independent study involved a study of basaltic dikes investigating possible fluid flow and phenocryst zonation. The student working on this study haa just completed Mineralogy and required only direction and consult on the research aspect of the project, not the instrumentation. Other science departments have found that students are more likely to undertake independent research after exposure to instrumentation in coursework. The number of independent studies could easily increase if CL were incorporated in the sedimentary petrology or paleontology classes, the classic areas in which CL is applied. An ore deposit class is another course to potentially use CL.

Overall I am not hearing a cry from the students that learning CL and its applications were a waste of time. As I fine-tune the assignments and work to include it in additional courses, the problems that students encounter with the instrumentation will lessen. Continued reinforcement of instrumentation in multiple science courses in a curriculum will very likely improve understanding of the uses and applications. Some day the use of CL by undergraduates could become as common place as the polarising light microscope.

Field-based Geologic Research

For a geologist or soil scientist, a soil is a naturally occurring body of roughly horizontal layers (horizons) located at the surface of the earth whose morphological, chemical, biological and morphological properties are altered from that of the original sediment or rock from which they have formed (parent material). Soils defined this way are distinct from those defined by an engineer who considers any unconsolidated natural material a 'soil'. A geologist's 'soil' is essentially the weathering rind of the Earth's surface. Thus soil properties are directly related to- and provide insights into- the processes and characteristics of the atmosphere, biosphere, hydrosphere and lithosphere.

Over the past four years, a central component of the lower level undergraduate soil science course at the University of North Carolina at Charlotte (UNC Charlotte) is the development of a soil geomorphology study that examines how soil properties vary in a landscape as a function of the characteristics of local environmental conditions. For the majority of students, this course is one of the first that they have taken beyond introductory geology. Numerous similar soil geomorphology studies exist for the western United States; however, very few have been undertaken in the temperate climates of the eastern seaboard, and fewer still in the Southeast. Thus, the project allows students to collect original data that can contribute to the field of study.

This type of inquiry based learning is a Recognised and recommended pedagogy that many find difficult to incorporate into lower level undergraduate course work. Nevertheless, numerous workers have demonstrated that such a teaching approach will improve students' overall science-related skills and understanding, including critical thinking and hypothesis development and testing. Furthermore, numerous studies have clearly demonstrated the benefits of field-based learning in introductory level geoscience courses. Commonly cited difficulties related to such field-based courses include expense, travel time and student-teacher ratios. Because soils (weathering profiles) are ubiquitous, however, lack of bedrock outcrop does not preclude completing a soil-based study, which can be developed on even the most urban of college campuses. Furthermore, basic soil field data are similar to those of geomorphology and sedimentology, so most geologists have the background knowledge required to develop such a course. Also, the repetitive nature of soil descriptions allows for working with 3-4 students at a time in a 20-30 student class, while other groups are engaged in activities that they are comfortable with. Finally, the skills and concepts learned during such fieldbased coursework have been demonstrated to be portable to other courses and to be professionally enriching. Here I describe a blueprint for the UNC Charlotte course that could be employed by almost any geoscience department as a 2nd tier introductory level fieldbased course.

Soil Geomorphology Concepts

At UNC Charlotte, concepts of pedogenesis and soil geomorphology are taught to students throughout the semester in separate lectures; however, a separate lecture is not necessary for the field-based component of the course that is described herein. Soil textbooks such as Birkeland (1999), Buol et al., (1997), Schaetzl (2005) and Brady and Weil (1999) are all good references for the following basic soil-geomorphology concepts that should provide a sound starting point for instruction.

A soil profile is described in a similar manner to a rock outcrop. Birkeland provides a manual for how to dig a soil pit and describe a soil profile that should be read and reviewed by the instructor and employed by students while in the field. This is a manageable summary of a significant body of widely available soil description literature. In particular Soil Survey Staff, 1993; and Schoeneberger et al., 2002 are both available for free on the web and are excellent resources for the instructor if necessary.

Description of a soil profile incorporates a description of secondary weathering features related to soil development (e.g. horizons, colours, soil structure) as well as any primary features inherent to the parent material such as sedimentary stratigraphy, grain size, or mineralogy. Beginning at the ground surface, soil horizons are typically vertically stratified with notable horizontal boundaries. Horizon boundaries mark changes in weathering characteristics and are located by marking significant changes in observable properties such as colour, structure (natural particle arrangement), texture (grain size), induration, root content and/ or visible pores. Horizon boundaries may or may not coincide with stratigraphie boundaries that mark different sedimentary or rock units. Once horizons are determined, a standard suite of soil properties is described for each horizon in a profile. Horizons are named according to an accepted nomenclature. A, B, C etc., depending on their properties. In a very general sense, A horizons are zones of leaching and of additions from the ground surface (e.g. dust in arid settings, organic material in humid settings). B horizons are zones of accumulation and in situ alteration (e.g. dominated by secondary salts in arid settings, iron oxides and clay in humid settings). C horizons are relatively non- weathered material in which the primary properties of the parent material dominate the horizon.

Once described, soil profile horizons and properties can be employed to better understand the soil forming environment unique to each soil exposure. Soil development is thought to proceed as a function of soil forming factors: climate, organisms, relief, parent material and time. All five factors influence the suite of chemical and physical weathering processes that serve to create any observed soil profile. Soils forming on toe slopes (an aspect of the 'relief factor), for example, will often exhibit redoximorphic features such as mottled colouring as a result of developing in a wetter soil climate than their backslope counterparts. Differences in microclimate might then be attributable to differences in the runoff and run on of these two landscape positions.

If four of the five soil forming factors are held constant for a group of soils, then the influence of the remaining factor on soil development can be determined. A soil chronosequence, for example, is a series of soils whose properties vary primarily as a function of the age of the exposure of the surface on which they are forming. Deposits or surfaces of similar age should share similar soil properties if other factors of soil formation are held constant. Young soils, for instance might be characterised by relatively thin horizons, and little evidence of insitute alteration of minerals. Well developed older soils, in contrast, would generally exhibit thicker horizons and profiles with relatively greater mass of secondary, translocated or precipitated minerals accumulated in the profile. A suite of inset fluvial terraces or alluvial fans are examples of settings in which a chronosequence might be developed. In a similar manner, a toposequence, climosequence or biosequence can also be examined. The field of study known as soil geomorphology has stemmed from the development of this soil forming factor concept and has contributed to many geologic studies throughout the world.

Soils are most widely employed by geologists as a tool for mapping and correlating geomorphic surfaces and surficial deposits in a variety of environments; however their application extends well beyond a relative age dating tool. Soils developing in deposits with numeric age control can be also be employed to determine a calibrated age for deposits that lack alternative age control. An examination of soil properties can provide insights into landscape evolution related to climate change or tectonics. Soil variability across a landscape typically can also be directly linked to the spatial variability of surface processes at work there. Finally, because weathering of geologic materials is so closely linked to moisture availability, soil development can be used to evaluate the hydraulic properties of geomorphic surfaces. Chronosequences, climosequences, toposequences, geosequences and biosequences are thus powerful vehicles for introducing and teaching a variety of geologic concepts and skills to students, for developing and testing hypotheses of varying complexity and for introducing students to field work in a simple, three-hour class throughout the course of a semester.

During the first two to three weeks of the UNC Charlotte course, students examine a stream cut bank on campus adjacent to the location of their future project. UNC Charlotte is located in the relatively urban Mecklenburg County in the Piedmont of North Carolina. The field site on the UNC Charlotte campus consists of a typical Piedmont 2nd order stream, Toby Creek, incised into a broad flood plain that includes small (1-2 m wide) levees and a number of level benches and low relief tributary alluvial fans projecting out from adjacent hillslopes. There, as a class, we make qualitative observations about the landscape and the sediments and the soils of the site. For example, I have students map the contact between the floodplain and the adjacent hillslope, using concepts such as changes in grathent, geologic and geomorphic units. Students are encouraged to make the connection and distinction between a landform and the underlying sediments which comprise it. Importantly, I help students to Recognise the differences between outcrop attributes that are the result of weathering versus inherent features of the parent material. In my experience, students have some difficulty at first with the concepts of "parent material" and of "soil" as a weathering feature (as opposed to the engineering concept of soil as any unconsolidated material). To have students compare unweathered sediment from the stream with weathered sediment in the cutbank serves to emphasize the difference between primary and secondary (weathering) characteristics. By the end of this exercise, students can Recognise and name basic landforms and geomorphic concepts.

Next, I divide the class into small groups and have each group claim a small (1 m) wide area of the cut bank that we have been examining. Students are asked to make observations about the exposure starting from ground surface. Armed with butter-knives they have brought from home, they draw horizontal lines where major visible or tangible changes occur in the vertical profile. Often students focus on colour, but texture and induration are also important properties to address. We regroup, and I explain that they have just determined the 'horizons' of the soil developing in the floodplain deposits of Toby Creek. We go as a group to each section of outcrop and students justify their boundaries to the class. Typically students have difficulty verbalising what is different about their chosen horizons beyond the property of colour. This exercise thus sets the stage for introducing the vocabulary associated with various soil properties such as texture and structure. At the end of this exercise, students can Recognise soil horizons, although they cannot yet name them.

One by one, I then demonstrate how the standard suite of soil properties is described. Students are provided with Birkeland which gives detailed instructions for describing each soil property. Additionally, I provide a modified soil description sheet. These description sheets are invaluable aids for keeping track of what must be described for each horizon. After each demonstration, students return to their own portion of the outcrop and describe the property for each horizon in their soil profile. Thus students repeat the process of description six to seven times. I split my time between different groups and assist as necessary. My emphasis in these first labs is focused on how to make objective, consistent observations about fieldderived data. Students are evaluated based on their participation (are they asking and answering questions; 50%) and on their field notebooks and worksheets (50%). Students receive points on the latter for completeness of their descriptions (are all properties described for all horizons), neatness (is everything readable and clear), and sketches (is there a scale, indicator of direction, etc.). At this point, I do not take off points for accuracy of their descriptions. I provide feedback after each lab, so that expectations are clear and improvements can be made for the next lab. After two labs, students are generally familiar with the mechanics of soil descriptions and are capable of making quantitative field observations.

Finally, I introduce students to the concepts of soil forming factors and soil chronosequences. Students typically find it intuitive that weathering will be affected by variables such as climate and vegetation; however concepts such as local microclimate typically require more attention. Discussing the soil forming factors approach allows review of concepts such as variables and constants and cause and effect in a geological setting. I provide examples of landscapes where chronosequences or toposequences have been examined in the past with a particular emphasis on fluvial terraces because they are a common landform in the Piedmont. The above information provides a segue to explain to the students that they will spend the remainder of the semester designing, and implementing their own soil geomorphology research project. I emphasize that the data that they collect will be unique, and thus will significantly contribute to our knowledge of soil geomorphology in this region.

As a single lab exercise, before we go to the specific field site, I supply students with a two foot-contour interval topographic map of an area of Toby Creek and indicate the field area where they will develop their chronosequence. In this portion of Toby Creek, there are several relatively flat concordant upland surfaces that students map. We review elements of topographic maps, and students convert the scale of the map from feet to metres (an exercise that seemingly cannot be repeated enough at this academic level).

The exercise ultimately requires students to develop a geomorphology map and a topographic profile of a transect across the field area. As a class, we qualitatively note areas of similar morphology (flat, vs. steep vs. very steep) and make hypotheses about origin of the landforms that the map is depicting (floodplain, possible terraces, tributary alluvial fans, and anthropogenically modified areas). I use the concept of 'terrace' and 'alluvial fan' loosely here as the scale is rather small, and I emphasize that by examining the map, we are developing hypotheses of stratigraphy and landform origin that can be tested in the field. I review the concept of a mapping unit and explain how a géomorphologie map differs from a geologic map. Together we define mapping units for the field area, and students then individually colour in their maps using provided possible mapping units and make a key to their units. Students are graded primarily on the accuracy of their topographic profiles and their ability to follow the provided 'rules' of geologic mapping. For example, all polygons must be closed or go off the map, all areas of the map must be assigned a mapping unit, and all maps must be neat with consistent and correct labelling. I next remind the class that the goal of the research project is to gain an understanding of how soil properties change as a function of soil forming factors in the setting of Toby Creek. Armed with this goal, their newly developed geologic maps and their understanding of soil forming factors, students develop hypotheses relating landforms and mapping units to soil properties. The list they develop typically includes the following: 1) if two areas on the map are assigned the same mapping unit (hypothesis), then the soils on those units should be similar (test of the hypothesis). 2) Soils developing on the floodplain should be less-developed than those on a landform mapped as a terrace. 3) If flat areas are terraces, then flat areas of progressively higher elevations should exhibit increasingly developed soils. Students use these hypotheses to choose and mark potential soil pit sites on their maps. Usually these pit locations consist of one to two pits on the floodplain, and three to four pits on two surfaces of different elevations which are office mapped as terraces.

We finally take these maps to the field. I spend a significant amount of time helping students orient themselves with respect to the topography and the units that they have mapped. Together, we find and examine the pre-defined pit locations and mapping units. We discuss map scale and topography as it relates to reality. In Toby Creek, small benches that students have often office mapped as terraces are identified in the field as tributary alluvial fans spilling out into die fioodplain from adjacent ephemeral streams. If necessary, we modify their choices of pit locations to allow for an analysis of both temporal and spatial variability of soil development (i.e. more than one pit on individual surfaces). Together we establish a naming scheme for all pits and for any samples collected from the pits.

Until this point, students have been working individually or in instructor-designated groups. I now ask the students to combine themselves into groups of 2-3. Typically there are 5-7 such groups each semester. I emphasize that working on teams is typical of research in any discipline and try to emphasize good teamwork skills. For example, students are told if there are conflicts or disagreements within groups that they be addressed as soon as they arise. To date, I have had very few issues with group work. Soil descriptions such as those executed in this project have been shown to foster good group-working skills.

Each group digs one soil pit from the list that we developed. Typically an entire class period is necessary for this task. If old pits are employed, they must be re-excavated. Pits are dug to about 1-1.5 m depth, taking care to preserve at least one single vertical face with undisturbed ground adjacent. Following procedures outlined in Birkeland (1999) and using the provided description sheet, each group describes the morphology of the pit that they have dug. Over the course of the following weeks each group describes all pits in the project. Typically a single description takes an entire lab period. Initially, each soil description is graded based on completeness of observations. For example, I will not deduct points if a student describes a loam as a silt loam, however I will take off points if texture is missing from their descriptions. In future classes, I become more concerned with consistency of observations and following of instructions. If a student describes the texture of a particular horizon as a loam, but assign it a consistence of very sticky; by the definition provided in the materials that I have given them, these do not coexist and thus points are subtracted. Each week, my expectations of their descriptions increase as they learn new concepts related to the soils that they are describing.

Results of the research are presented and discussed at the 'Soil Science Class Research Symposium' held at die end of the semester. This presentation constitutes a significant portion of the final grade for the course and is a substitute for a final exam. I often invite other faculty and students to attend this event. Each group develops and executes a 30 minute GSA-style presentation with each member of the group taking responsibility for different portions of the talk.

Each group presents data from all soil pits. Students are required to incorporate concepts and terminology learned throughout the semester into the presentations. I challenge students to avoid haphazardly presenting all data collected, and to focus their presentations on one or more of the hypotheses that were developed earlier in the class. I encourage them to focus on the portion of the research or the observations that they found most interesting. I urge them to be creative in terms of data presentation so that, visually, the data are clear and unambiguous. As part of their final presentations, students are asked to consolidate their field descriptions into meaningful diagrams or graphs. I encourage students to translate soil description data into numerical semi-quantitative indices so that they can be more easily compared. Students may create their own indices or they may choose to use those available in the literature, such as the classic profile development index of Harden (1991). In the Harden index, for example, soil properties are compared to those of the parent material (in this case unweathered fioodplain sediment), and deviations from the parent material are assigned a number. Increasingly higher numbers indicate increasingly more developed soils. Creating and using such indices provides good experience in translating field data into a semi-quantitative record. Presentations must contain the following elements: 1) An introduction and justification of why the research is important 2) A clearly stated hypothesis or goal 3) An explanation of the methods used to test they hypothesis and achieve their goals 4) Results presented in visually clear graphs or diagrams and 5) Interpretation and conclusions in the context of their previously stated hypotheses and goals. Students are provided with a written description of my grading scheme for their final project presentations:

1) Content 33%. Is there a clear problem and/or justification of the topic presented from the beginning? Does the presentation cover the stated topic? Is there originality or creativity in the interpretation or presentation of the data? Do students correctly incorporate concepts and terminology learned in class in their presentations?
2) Organisation 33%. Is there a logical flow of thought in the presentation? Does the authence know from the beginning what is going to be discussed?
3) Presentation 33%. Do the speakers speak clearly? Are visuals clear and readable?

As a final portion of the grade for the class, each student in the class is also required to ask one or more questions of another group during the discussion time. After all presentations are complete, we summarise the main similarities and differences between groups and put the overall project back in the context of soil geomorphology research in the Southeastern United States.

In 2005, students mapped two tributary alluvial fan surfaces and two 'upland surfaces' in their designated field area. Field relationships indicated that one alluvial fan was inset into another so a relative age relationship was established, with Alluvial Fan 1 hypothesized to be older than Alluvial Fan 2. It was hypothesized that the upland surfaces were related to long-term fluvial incision, and thus the higher of the two surfaces (Upland Surface 1) was hypothesized to be older. For logistical reasons, we chose to focus our study on the tributary alluvial fans and the fioodplain. Five pits were excavated and described by each group of students. Pit #1 was located on the fioodplain. Pits #2 & #3 were located on Alluvial Fan 2, whose surface is approximately two metres above that of the fioodplain. Pits #4 & #5 were located on Alluvial Fan 1, whose surface sits approximately 4 metres above the fioodplain.

These are typical of any year's class in that they range from purely illustrative to quantitative. The sophistication of the visuals also is variable, but are accepted and allow less-computer savvy students to feel on equal footing. The following is a combined summary of the results presented by students for the 2005 project. The detail of the results is illustrative of the complexity of the ideas and conclusions that students are able to make using the data that they have collected. With only minimal change in emphasis of data collection and analysis, the same project could have highlighted other geologic disciplines such as geomorphology, sedimentology or mineralogy.

For the 2005 project, there were notable differences between soils on different surfaces and between soils on the same surface. The fioodplain deposits expectedly exhibited relatively weak overall soil development with A-BwC horizonation with a buried soil at depth overlying undisturbed fluvial stratigraphy. Students noted that the surface soil in this pit exhibited slightly redder colours and more structure than the cutbank exposure they had previously examined. Also, the soil in Pit 1 was lacking the unweathered sand deposit from a 2004 flood which we had also observed in the cut-bank exposure. These observations led students to examine the location of the pit in more detail and to Recognise that the area in question was potentially bypassed from flooding due to the presence of a crevasse splay in the nearby levee. I used this and other similar experiences to emphasize to students that we can often learn more when a hypothesis that we propose (in this case, soils developing on a single geomorphic surface should exhibit similar characteristics) is proved false.

The soils exposed by Pits 2 & 3 on the lower alluvial fan exhibited similar characteristics. Both soils are developing in homogeneous, well-sorted grussy-sand that is consistent with a tributary alluvial-fan origin for this surface. The soils exhibited diffuse relatively dark A horizons which extend to approximately 20 cm to 40 cm depth. These two soils also exhibit approximately 20-40 cm thick oxidised B-horizons with few if any clay films evident. Colours at the bottom of these pits were approaching tli.it of the fioodplain sediment.

Pits 4 and 5 shared both similarities and differences. Pit 4 and the upper horizons of Pit 5 were characterised by similar well-sorted sand as pits 2&3. The A horizons for these pits were much thinner (only 6-10 cm in depth) than those of Pits 2& 3 and were also much more concentrated in organic material as evidenced by darker colours. These A horizons overlie a discrete Bt horizon of 10-15 cm thickness with well developed clay films developing in the sandy parent materials. In Pit 4, this Bt horizon graded into a well-oxidised B horizon with occasional clay films present. Underlying a relatively thin Bt horizon in Pit 5, however, was extremely weathered, clay-rich saprolite with extensive mottling, gleying, and clay films. The overall soil development of this portion of Pit 5 was significantly greater than the other four soils examined. This difference in soil development is reflected in a high overall Profile Development Index.

In the case of the 2005 study, most groups Recognised the overall increase in soil development in deposits of increasingly higher elevation. There were notable exceptions to this trend, however, and these were discussed by many groups in their presentations. For example, many students noted the presence of Mn nodules in all pits regardless of age. Thus, students were able to conclude that the presence of Mn is not a good indicator of soil age. The group that calculated clay film indices noted an unpredicted relative decrease in clay films from the floodplain soil to the soils on the lower alluvial fan. It was suggested that the decrease was due to the fact that the buried soil in Pit 1 was used as a parent material reference, likely not an appropriate choice for their index. All groups noted that Pit 5 was the only pit prone to filling up with water after a rainfall. Many groups concluded correctly that the high clay content of this pit is contributing to periods of sustained saturation, and thus red-ox conditions, as evidenced by the extensive mottling observed there.

Some groups noted the differences in A horizon morphology between Pits 2&3 and Pits 4&5 and after some prompting were able to draw some conclusions. The relatively high permeability of the sandy matrix in Pits 2&3 permitted dark organic material to be transported to deeper portions of the soil profile, thus resulting in relatively diffuse, thick A horizons. It was suggested that this transportation was not possible in the soils of Pits 4&5 because sufficient clay had accumulated to retard infiltration into these soils. Thus the soil development of the B horizon resulted in a change in the development of the A horizon. A final interesting conclusion from the 2005 project was the recognition that deposition of tributary alluvial fans in the Toby Creek drainage is episodic, as is evidenced by fans with different degrees of soil development. Pit 5 provided evidence that the thickness of these fans is variable and that an older, more weathered landscape is preserved beneath them.

Numerous aspects of the soil geomorphology project have produced positive learning results and experience for students. First, I have found, as have others that putting the students' work in the context of a topical research problem engages students and encourages care in data collection and critical thinking in a way that a simple assignment does not. Each individual exercise or soil description is given meaning beyond the mechanics of the assignment itself. The fact that students were able to take simple observations, such as noting the presence of Mn nodules, and discuss them in the context of broader hypotheses, demonstrates the success of the project in capturing the students' attention and creativity. The quality and depth of presentations and proceeding discussions demonstrated students newly acquired abilities to synthesize their own field data in the context of new pedologie concepts and their own hypotheses. By the end of the project, students are employing geologic observation skills, vocabulary and concepts with a sometimes surprising fluency.

The teamwork component of the course is an overall positive experience for students. I find that working in teams inspires students to assist one anther, creates a more continuous flow of ideas and helps with the tedium associated with describing sediment and soil properties. I find that group work inspires students to go beyond what is expected of them. For example, students are only required to describe each soil pit in the project. Most groups, however, returned to the field area after class to collect additional data or documentation. For example in 2005, most groups took photographs of the pits and surrounding area, and one group took microphotographs of samples. These types of additional efforts provide further proof that the project has promoted scientific inquiry and made students feel invested in the outcome of their work. The majority of difficulties in group work have come from the preparation of the final presentation. With emphasis on work equality when the assignment is made, however, I can generally reduce scenarios where one person in the group does the majority of the work.

Although digging soil pits is arduous exercise, I find that students feel more 'attached' to their pit once it is excavated than they do if someone else dug it or it was a natural exposure. Such sweat equity investment unavoidably leads to curiosity and exploration. As students dig, they inevitably note differences in soil properties such as induration and colour. These differences seem to pique their curiosity (or perhaps provide an excuse to stop digging) and to generate questions, therefore digging is beneficial and not a waste of class time.

The repetitive nature of the soil descriptions also serves students well. All students make noticeable improvements in the accuracy and precision of their observations over the course of this data collection portion of the exercise. By the last pit, students feel confident in their ability to describe sediment, rock and soil. By the end of the course, students feel comfortable and are successful in presenting results of their data collection during the 'Seminar'. The excitement of the seminar comes, however, when students discuss their interpretations of the data after all presentations have been made. Each student is required to ask 3-4 questions of the different groups as a requirement of the seminar. The groups typically share similar data-sets and are often 'amazed' that that most groups independently made the same observations that they did. Although different groups tend to focus on a variety of differences between soil pits, because all students have collected data from the same sites, they quickly become engaged in the question and answer section and the discussion.

Students' overall feedback regarding their semester long project has been positive. On more than one occasion, students have felt sufficiently invested in their work to present their results at the UNC Charlotte- and/or the North Carolina State- Undergraduate Research Symposium. Students' independent efforts to develop these posters and present them at conferences speak to their enthusiasm about the research that they have completed. Other students have indicated to me that the knowledge and skills that they gained during the soil chronosequence project aided them in gaining employment with geotechnical and environmental consulting firms. In general, the soil geomorphology project in the UNC Charlotte Soil Science course has been a positive component of our geosciences curriculum.

As other studies have demonstrated, it has been my experience that a field-based research project can provide a good first comprehensive exposure to geological field research. Soil development is closely linked to the properties of the environment in which soils are forming. Thus in a soil-based project, students are required to make sedimentological, geomorphological, hyrological and ecological observations in order to interpret the soil data that they collect. This type of soil analysis far exceeds the coverage of Soil Taxonomy that is included in most introductory geology textbooks. Soils developing in floodplain and adjacent environments in the piedmont of North Carolina are influenced by the local hydraulic and geomorphic conditions of the landscape in which they are forming as well as by the temporal changes in the physical properties of the parent materials. It is an important lesson for students to Recognise that soil properties can provide enormous information about the landscape in which they are forming. Concepts related to such a project are universal to field based research and thus provide a sound first exposure to geological field work and practices.

9

Technology, Accuracy and Scientific Thought

Geologists generally recognise field camp as the time when undergraduates make (or begin to make) the key transition from student to scientist. Field camp is a capstone, integrating separate courses such as petrology, stratigraphy and structural geology into a single experience. This integrative experience is part naturalistic description of the physical world, and part scientific interpretation of its long-term history. Field camp is a unique opportunity to study the real world, while reinforcing classroom experience and building student confidence and interest. Despite these benefits, the number of field camp courses being offered nationally has been steadily declining for some years. Field camps are expensive, and generally do not recover their costs through tuition. In times of shrinking budgets, many programs are forced to fight for their field camps, or to simply eliminate and outsource them. It is now, more than ever, crucial for us as geologists and as educators to justify the continued effort and expenses associated with field education.

The purpose of this ethnographic study is to support field education by a narrating and interpreting the mental lives and constructed realities of a sample of undergraduate field camp students. Many workers have described and analysed field education. The importance of field mapping from the perspective of a senior geoscientist. He considers field relationships to be the "ground truth for all earth science investigations". The importance of field-based learning and teaching as part of the broader geoscience training of K-12 science teachers. They argue that geoscience departments directly contribute to and merge with their universities' missions by training preservice teachers in earth sciences. These workers then argue that for the general public, a strong background in the earth sciences (and by definition, this includes field-based learning) is crucial to coping with public welfare issues. Elkins and Elkins (2007) conducted an outcomes-assessment study comparing field and classroom learning experiences. Their study provides statistical and empirical data as evidence of the benefits of field education. Riggs et al. (2009) conducted a study of field camp learning as a function of navigation and movement.

In this study, students carried global positioning system (GPS) transmitters that recorded their movements, which were overlain onto a geographic information system (GIS) grid. Riggs et al. (2009) then produced GIS polygons that described each student's navigation choices over the landscape throughout their mapping efforts. These workers argue that learning by moving through a landscape has no classroom analogue and is, therefore, an implicit justification for field education. Riggs et al. end their discussion with a call for the use of qualitative methods in order to more deeply understand student navigation choices, their behaviour in the field, and the nature of their field-situated learning. This ethnographic study is a response to their call is through formal, phenomenological exploration of the "essence" of student field experience. This essence, and how it shapes field learning in the geosciences, has not been well studied. By "essence" I mean such things as day-to-day life in the field for students, their interpersonal interactions and, most importantly, how students construct reality, both on a private, individual basis, and as part of an interactive class group. The process of understanding these phenomena is distinctly different than measuring parametres such as what students know, identifying an improvement in skills or even assessing the impact of affective phenomena. Skills improvement, "knowing" and cognitive processes, in general, are in large part the end results of internal processes such as lived experiences and the personal (private) construction of reality. This is not to say that reality is only constructed through private processes. Given the highly social and interactive nature of field camp classes, the construction of reality is also dialectical and interactive in nature.

Petcovic et al. (2009, 2007) explored skill acquisition and cognitive processes in their studies of novice-toexpert progress in geologic mapping. These workers tracked field mapping teams with GPS, followed up with interviews and systematically compared individual maps to evaluate them for completeness and accuracy. Students also recorded audio logs during their mapping activity. Petcovic et al. (2009) concluded that the metacognitive skills of synthesis, hypothesis testing and ongoing revision distinguished novices from experts. These workers focus on the metacognitive differences between novices and students with greater potential to become experts. How are these differences developed? What are the internal processes- such as the private construction of reality- that lead to metacognitive sophistication?

Many workers consider these internal processes to reside in the affective-domain of learning. No matter how we decide to locate or describe these processes, they remain critical in shaping the outcomes we measure. Yet internal processes resist quantification and measurement. They are best studied not through quantitative instruments, but rather by immersed, painstaking observation and documentation to produce a compelling narrative. This narrative is a detailed story of the processes and benefits of field camp learning, presented in the ethnographic tradition of inquiry.

Ethnography is the traditional domain of anthropologists who immerse themselves in a "culturesharing group" in order to study that group. Much of the time, entirely different societies are the focus of ethnographic study. But the students enrolled in a field camp geologic mapping course are also a culture -sharing group, by virtue of their common experiences of travelling to the mapping site, negotiating the landscape, comparing and compiling joint maps, interacting with instructors, and living together under primitive conditions. Students in field camp classes also demonstrate shared patterns of observable behaviour. The culture sharing in a field camp class is independent of the students' demographics, but is rather an expression of shared experience and behaviour. As a culture-sharing group, a field camp class is appropriate for ethnographic study.

In the summer of 2009 we conducted just such an ethnographic study. My goals were to document and understand how the enrolled students constructed knowledge, perceived reality and coped with novel field experiences. We expected multiple themes ("results") to emerge that would describe the field learning process, which in fact did happen. In this chapter we describe just one set of "results" and implications, which is the technology dependence and construction of scientific reality that I observed among field camp students who were assigned to measure an interval of stratigraphie section. Before I present the methods and results of this study, it is important to note that ethnography is not a strictly empirical process, such as that typically practiced by traditional geoscientists. In order to provide a context in which to understand how I obtained my "results," what follows is a brief discussion contrasting the ethnographic method with our more familiar physical science methods.

Many geoscientists have legitimate concerns about the nature and appropriateness of non-numerical methodologies of inquiry, such as ethnography. However, some educational problems in the geosciences require qualitative approaches, because they revolve entirely around the personal realities, perceived truths and lived experiences of human beings. These variables resist quantification and their study is unfamiliar to many geoscientists. Furthermore, the ethnographic method is markedly different than the familiar empirical method. To be explicit: this study is not an experimental study. It is an ethnographic study of learning in the geosciences. In typical physical science praxis, a geoscientist (who can also be an educational researcher) observes a phenomenon, formulates a hypothesis, and then goes out into the field to collect data. The research process is hypothesis-initiated, and the basic order here is:

Observe devise hypothesis- collect data revise/reject hypothesis.

To use the ethnographic method, the geoscientist (an educational researcher) goes out into the field to generate data, figures out what she or he observed, and then narrates those observations and extracts meaning by describing a set of emergent themes. So the basic order here is:

Generate data synthesize observations extract meaning In the ethnographic approach to educational problems, the data are non-numerical, the approach is nonlinear and recursive, the researcher is not independent of the phenomenon, and the interpretations are not extrapolated to the general population. Ethnography is not hypothesis-initiated; rather, the hypothesis is explored after the observations are synthesized - this is the extracted meaning. In fact, this alternative methodological order occasionally shows up in the geosciences. A specific volcanic eruption, a meteorite impact, and a severe flood are all examples of ephemeral, non-reproducible phenomena that must be observed, measured or sampled right when they happen. The attending geoscientists form and evaluate hypotheses on the fly, or after the fact.

The purpose of a hypothesis is to explain an observed phenomenon in a generalisable way. A good hypothesis should be extrapolated from a sample size N to an entire population. In ethnographic study, the purpose is not to explain or represent an "entire universe" but rather to provide a vivid, illuminating "flavour" of a particular phenomenon. What I found in the current study may be pervasive and occur everywhere. On the other hand, that may not be true. The usefulness of this study is not in how it predicts exactly what will happen when students measure section (it does not), but rather in how it sheds light on, or varies from, the experiences, past or future, of other students who measure section, and faculty who teach them to do so. It is true that this is a generalisation towards a larger population, but not to the extent that a scientific hypothesis generalises. It is possible to extract meaning from a study even though it does not model an entire population. This meaning comes not from the verification of a hypothesis but rather the researcher's generation of data, his or her identification of patterned regularities and the development of themes in those data, and how those themes are interpreted and shared.

The issues of "validity" and "reliability" do not apply to ethnographic data; these are parametres of numerical inquiries, which ethnography is not. However, ethnographic data are subject to analyses of trustworthiness and reliability. Ethnographers must establish that their data generation strategies are of the highest possible quality, and that their interpretations of the data are correct. In this study, I have worked to establish trustworthiness and reliability through two means. First, I have provided excerpted data to support my interpretations. One of the inherent risks of presenting interpretations of qualitative data is that the process may appear to be "black-box" in nature; that is, opaque, mysterious and difficult to agree or disagree with. In this chapter, I provide the raw data of statements and actions, and my subsequent interpretations of them. Therefore it is possible to see how I constructed the latter from the former. This process establishes trustworthiness. The second method I used focuses on reliability. Ethnographers will often "triangulate" their interpretations in consultation with other experts. In such a process, a worker has one or more other experts to verify that the emergent themes are consistent with the data. In this study, I triangulated my interpretations with both a cultural anthropologist and a geoscientist to establish inters -coder agreement.

It is worth noting that qualitative inquiry in general, and extracted meanings (themes) in particular, can contribute to the development of a hypothesis in a more traditional (i.e., quantitative or experimental) study; qualitative inquiries often seed quantitative investigations. We can think of the process of extracting meanings as the "methods" of an ethnographic study. I present this study in a manner different from the familiar format of Methods; Results; Discussion. This format is characteristic of experimental, hypothesis initated reports, which this study is not. Instead I blend a more traditionally "scientific" style with the format prescribed by Wolcott (1994) for presenting ethnographic data as follows: Methodology; Method; Narrative Description; Analysis; Interpretation and Implications.

The term "methodology" as we use it means the theoretical frameworks within which I approached and conducted this research, as well as my place in it as the researcher. The term "method" describes what I actually did. For example, interviews and focus groups are two methods, which will be used differently depending on the researcher's methodology. An ethnographer will extract themes; a phenomenographer will compare transcripts for variant experiences; a policy analyst will identify the impacts of a particular policy (or lack thereof). In this study, my methodology was hermeneutic ethnography and phenomenology; my methods included participant observations and interviews.

This study is informed by phenomenology because I endeavoured to understand how students construct reality and truth in what Orion (1993) calls novel spaces. What I considered relevant data are personal accounts and narratives, non-verbal behaviours, student-to-student interactions, individual choices, strategies and expressed attitudes. The ontological properties, i.e., basic realities, I investigated included people, people as social actors, emotion, memory, consciousness, understandings and interpretations, ideas and perceptions, attitudes, beliefs and belief systems. These data and ontological properties form and reside in communicated truths, which are not subject to empirical verification or numerical validity analyses.

The term "hermeneutic" is descriptive of my role as a researcher-practitioner. I sought to document field learning, but as an educator, I would use what I found to enhance and inform my own teaching. It was impossible for me to separate my identity as a teacher from the learning I was observing. As a result, I was constantly moving back and forth between my roles as an on-site researcher and a practicing teacher. This movement is labelled "hermeneutics". A hermeneutic approach seeks to understand a larger process through the understanding of smaller parts of that process, which in turn requires an understanding of that greater process itself. Further, a hermeneutic approach seeks understanding through a communicated truth, which itself is not verifiable. To understand something is to participate in it, not something to be relayed by an outside observer, nor to discover empirically.

For this study, an ethnographic approach was preferable to either a case study or a grounded theory approach. A case study would be appropriate (but not limited to) a situation where a single person's experience was being studied, especially if that person was in a particularly novel situation (e.g., she or he was mobilityimpaired, or in a virtual field environment). A grounded theory approach is suitable for workers who are seeking to define a consistent, generalisable model of a process. I attended a field camp course in 2009 from start to finish, accompanying and observing student groups as they set up and broke down campsites, participated in lectures and discussions, mapped field areas, compiled reports, and measured and documented a stratigraphie section. I obtained formal, informed consent from the student participants, and introduced myself to them at their orientation meeting. Once in the field, I occasionally placed myself in a more distanced position as an "observer."

However, most of the time I was fully integrated into camp life, participating in chores and social activities. I conducted and transcribed formal structured interviews, and I engaged in unstructured, conversational interviews. These latter interviews were in some cases completely unstructured, and in other cases were guided by a "mental checklist" of items that I compiled on-site. I coded transcripts and my observational notes in a simple serial indexing process, grouping interview responses and observed behaviours into basic categories. This produced data "packets" that allowed me to perceive themes across different transcripts and different behavioural observations. For example, if in one recorded interaction I flag "technology dependence" in a serial indexing process, I then could lift all comments and behaviours under that theme from one transcript to compare it to the same flag in another transcript or observation. I then processed the extracted themes into mechanical arguments, presented in the section "Analysis".

Description

In the presentation of ethnographic data, the Description is a "straightforward [account] of settings and events", so that the reader has a vivid, detailed and interesting story that serves as a foundation for the inquiry. In the description, the reader sees what the researcher saw, through the researcher's eyes. The description should not be detached or impersonal. It can be chronological, it can go through a list of subjects in a particular order or it may be written in a day-in-the-life format. I have outlined my description chronologically, with occasional subordinated descriptions of characteristics, personalities and/ or behaviours. I begin my description with a discussion of the class itself.

The field camp course we was immersed in is offered on a yearly basis by a public university ("University") located in the western United States. In order to maintain the confidentiality of the students, staff and faculty, I do not divulge the university or specific, identifying details about the course or the geology of its field areas. The course is team taught by senior faculty members (Professor X and Professor Y), and staffed with multiple graduate teaching assistants. Students are transported to three remote sites (labelled here as Sites A, B, C) for up to ten days at a time over the length of the course.

They do not spend equal amounts of time at each site; the time spent at each site is progressively longer going from A to B to C. About ten of the students, or one-third, were from the local university; the rest were from other institutions that require field camp, but do not offer it.

The student population represented a total of eight out-of-state institutions and one in-state institution, in addition to the host University. In this course, students and staff camp under primitive conditions without bathing or permanent latrine facilities. Kitchen facilities include portable propane stoves, but no running water. Students and staff provide and cook their own meals. All field areas contain rugged topography, with up to 150 metres of relief in some mapping areas. Structural deformation in the mapping areas ranges from moderate to extreme.

Students are given the primary task of compiling geologic maps of bedrock and surficial features. They also report on each area's regional geologic setting and tectonic history. They examine aerial photographs, collect structural data, assemble lithologie descriptions and, at the third mapping location, Site C, they measure and describe a stratigraphie section.

Students measure this section before they begin mapping, as a means of becoming familiar with the local rocks. The section they measured is located several kilometres from their mapping area. The rocks are part of a sequence of late Phanerozoic marine sediments that are very well exposed along several roadcuts. The package contains a total of nine discrete units, which have a moderate dip but are otherwise undeformed. However, these same units are intensely deformed in the mapping area. In this chapter, we describe the events that took place during the first full day of measuring section at this location, which was Day 12 of the course overall. Professor X led this activity; Professor Y was not yet present, but was scheduled to arrive within two days' time.

Before we go any further, it is worthwhile to go back in time to the second day of the field camp class, in order to set the stage for the events on Day 12. On Day 2 the class made its first visit to Site A, an area of moderately deformed sedimentary rocks and rugged topography. The students were each given a 1:6000 scale topographic map and instructed to locate themselves on their map. They were not given GPS units, but were told to use ridgelines and other topographic features to find their location. It was apparent to me that the instructors valued the skill of doing field tasks like this "by hand," versus relying on technology. I followed a small group of students, which included an individual named Joey-Ramon (Pseudonym. Assigned pseudonyms do not reflect the ethnicity of the participants). When his group split off to locate themselves, Joey-Ramon had this immediate comment:

This is retarded; we have GPS and could figure this out in three seconds. What's the point of this? On Day 9, while in the field at Site B, we had the opportunity to interview Joey-Ramon and his mapping partner. We asked Joey-Ramon about his sentiments above. In fact, we read his comment back to him verbatim. Below are excerpts from his response, edited for redundancy and clarity:

Yeah, my sentiments are still mostly the same.We understand more or less now what [Professor X is] trying to do, because we think it's - it's kind of an art, reading a topographic map, there's not a lot of people that can do it and then having taken structure, where we've gone out in the field and just located ourselves through GPS, it didn't seem like we needed to do that anymore. It's kind of a luxury that we have [GPS], living in our day and age as opposed to somebody - way back in [Professor X's] time, they didn't have that opportunity, so they had to do it based on a topographic map. So I have like a new found respect for it, I think it's kind of like an oldschool thing to do, for lack of a better term, and something that our predecessors had to do and I think we do things differently now. I think in this day and age, we've grown up with technology, and we've just learned to use it but I've never questioned it.

Returning to the events of Day 12 at Site C, the students were introduced to the lithologie units that they would find in their mapping area. Professor X began the day's discussion by asking students to describe the basal unit of the section, a thickly bedded marine sandstone unit. Most students looked at their handouts instead of the rocks. They called out formation names based on what they could visually correlate with the information on their handouts.

One group moved off to take a strike and dip measurement. The teaching assistants then prompted the students to look at the rocks first, before attempting to name the unit. Within a few minutes, consensus was reached on the name and properties of this unit, which they had in fact seen at their first mapping site! Professor X led a discussion about paleogeography, asking the students to consider whether this location represents a shallower or deeper marine setting than what they mapped at Site A. A vocal minority of students decided this location was shallower, based on what they could remember of the grain sizes they saw nine days prior.

The rest of the group remained passive. Professor X responded by constructing a rough map in the dirt parking area. He led a Socratic discussion of the regional paleogeography, and eventually the students realised that by placing the mapping areas in context with paleoshorelines, the present setting was actually deeper. Professor X then walked the group over to a contact between the top of the marine sandstone and a shale unit. He instructed the students to follow the shale upsection. As they did, he led them through a Socratic discussion of transgression and regression. Professor X then gathered the group in the parking area in order to give specific instructions and outline expectations for the stratigraphie measurement activity. Professor X is a veteran teacher who is highly skilled in active-listening techniques. He has been the principal instructor for this course for over two decades. He speaks deliberately and is not glib, and he converses with students in a respectful manner, with gentle humour. He is highly tuned to non-verbal behaviour among students, gauging their understanding and comfort very quickly and without words. Throughout the course, students demonstrated a consistent, willing enthusiasm to internalise his advice and follow his instructions. At this point on this day, he distributed 1.5 m Jacob staves, and instructed the students in how to install a Brunton compass onto each staff. Professor X then instructed students how to measure thickness by orienting the staff perpendicular to dip and walking along the road in an upsection direction. This technique was subsequently referred to as "jaking." Professor X repeated the explanation, and physically demonstrated the technique. Most students were not familiar with this technique, and they were enthusiastic about trying it out. However, Professor X was explicit in his directions; he wanted students to use the staff as a ruler ("ruling"), when appropriate. He made it clear that a majority of measurements were to be made by ruling. He wanted them to climb the rocks, and physically lay the staff on the rocks perpendicular to dip as they moved upsection. Professor X gave this instruction once before he taught them the jaking technique, he repeated it during the demonstration, and he repeated it once again as he broke them up to conduct the exercise: "Use your staff as a ruler!" I initially followed two students, Chuck and Sid.

These two, along with most students, expressed a great deal of initial confusion and a lack of confidence as they started the task. Chuck and Sid began by taking a strike and dip measurement. They then oriented their staff perpendicular to dip and began jaking the section. As they moved upsection, they did not hold the orientation of their staff consistently. As a result, most of their measurements of the basal unit were oblique to, not perpendicular, to its dip. When they reached the contact between this sandstone unit and the overlying shale, they compared their determined thickness with that of two other groups. Sid and Chuck were dismayed to find that their result varied substantially with other groups. I then heard Professor X in the background, earnestly redirecting students to rule instead of jaking the section. Professor X was careful not to show his mild frustration, but I was in a position to observe it. Despite the detail of his instruction, there existed among the students great deal of uncertainty about this task. I moved over to Professor X and away from the students.

> *"Is this normal?" I quietly asked him, while gesturing towards milling groups of students.*
>
> *"What?" He scowled, visibly annoyed.*
>
> *"This," I responded, gesturing to the students.*
>
> *"You mean mass confusion?" Professor X shook his head.*
>
> *"Yeah," I responded, "And jaking, instead of using it as a ruler."*
>
> *"Every year," Professor X shook his head. "It's like they don't believe me. They... they don't believe me. Every year, this happens." [Emphasis his.]*

We wandered away from Professor X to consider his statements. So far in the course, I had observed students carefully seeking and internalising faculty advice and instruction. In fact, students were more inclined to seek external validation than to trust their own decisions. This was the first time that I had observed a mass disregard of either advice or instructions, and I found it frankly puzzling. I was not willing to concede that in this one instance, they arbitrarily chose, en masse, not to "believe" Professor X. We decided to directly ask the students why they chose the technique they were using to measure the section. I moved through the area to converse with nineteen students distributed in nine different mapping groups. Each group I spoke with was jaking, and I asked them why they chose to jake instead of choosing to rule the section. I posed the same question in the same manner to each group. Their responses are summarised below, either by individual or grouped pseudonyms, in order. I did not audio- or video-record these conversations, and in most cases I was not able to write down full verbatim responses.

Analysis

In an ethnographic study, the "Analysis" section describes the researcher's transformation of data into an argument. In this chapter, the student responses to my questions are raw data. I describe what I flagged (observed as recurrent and significant) and how I interpreted these flags to be descriptive of meaningful ideas and processes related to student learning in the field, i.e., the argument. For further reference, Libarkin (2005) discusses qualitative analysis relevant to the geosciences, and the anthropologists Ryan and Bernard (2003; 2000) provide in-depth discussions on thematic analysis in ethnography. Two emergent themes are readily flagged in these responses. First, the notion of finding a physically easier path, literally and figuratively, is prominent in these responses. I suggest that there is nothing at all new or interesting about students (or people in general) wanting to do things in the easiest way possible. Therefore, I have ignored these flags, and the emergent theme they raise of "the path of least resistance in fieldwork."

The second emergent theme is that of a desire to be correct, through being accurate. Unpacking their statements and behaviours further, it becomes apparent that the students feel they attain this accuracy through applying technology, in the form of the compass - which they discussed frequently. A Brunton compass is not computerised technology, but it is technology nonetheless, in the same way that wrenches, screwdrivers and a carpenter's level are technology. These tools are a higher level of technology than that of stone tools, and lower than that of GFS satellites and receivers. All levels of the technology hierarchy may be used to obtain information about reality, and therefore can be perceived as having some role in "truth."

Educators and scientists tend to hold the general view that more technology is better. This argument is supported by 1) the growth of "online" education across STEM disciplines; 2) the emphasis placed on cyberlearning initiatives by the National Science Foundation (NSF); 3) initiatives to increase computerised instruction in K-12 classrooms; 4) the role of technology in scientific revolutions; and 5) the continuing funding of the development of instrumentation by federal granting agencies such as the NSF and the National Institutes for Health. Strong statements about the value of technology exist in the NSF's Mission and Core Values statement, as part of its Strategic Plan.

Ostensibly, better technology improves our understanding of reality; otherwise, why would we invest so much in its development? So it seems that scientists, politicians and teachers agree that more technology is better. This ethos permeates our lived experiences, and it should be no surprise that it would manifest among our students, and that I would easily observe it in the field. I observed a desire on the students' parts to be accurate, and their attempts to actualise this through applying the technology of the Brunton compass. I also observed that they were, in large part, using the technology incorrectly. Their Jacob staves were often not oriented perpendicular to dip, and yet, it did not occur to them that they were using their tools incorrectly.

They had a strike and dip, they matched their measured dip with the inclinometre on their staffs, and that was that. The numbers gave them certainty, in the way that visual ruling could not.

These students had low levels of confidence in their "low- tech" field skills, and they did not trust their visual observations of spatial orientations. They relied on available technology to compensate for their lack of confidence, as well as their reticence to exercise their skill in "low-tech" measurement. The student Skye summed up this concept in her comment, "This is why we have compasses - accuracy makes science." That is, the ability to assign a numerical value to a behaviour- in this case, how to hold the Jacob staff- made the process more reflective of scientific reality. The term she used was "accuracy," but what she appears to have meant was "precision." She interchanged the two concepts in her mind, as did all the students.

However, we argue that the two concepts have important differences. "Precision" refers to measurement; that is, how many significant figures, how much resolution, how many pixels, and so forth. "Accuracy," on the other hand, refers to reality; that is, something is located here and not there, it is 12 cm and not 10 cm thick, it is fluvial and not deltaic. In this latter example, the differences between interpreting an environment as fluvial or deltaic rely on observation and measurement. As geologists, we want an appropriate level of precision in our instruments, and we also want to be accurate; not "as accurate as possible," but accurate in a way that reflects scientific and physical reality. Sometimes we get only the precision part right. For example, we can conduct very precise measurements and observations, but incorrectly interpret the environment. We had a high degree of precision, but we were wrong- that is, we were not accurate. The number of significant figures does not prescribe reality.

Joey-Ramon's discussion of reading topographic maps with and without GPS contributes to the theme of technology dependence among students. For him, reading the map without the GPS is clearly an art, versus a science! To him, this skill is simply a classroom abstraction, a training activity conducted by his "old school" professor. Modern science is conducted with technology, which Joey -Ramón grew up with and never questioned. Technology renders the old skill set obsolete. Why is it obsolete? JoeyRamon does not value the technology for its own sake; he values it because he feels it is better descriptive of reality. GPS coordinates give him a more precise location on his map.

Therefore, he can develop a more accurate picture of the geology of the area. One question I cannot answer is why Professor X shared the jaking technique to begin with. This is troubling, because if he really did not want his students to do it, he would not have demonstrated it. Perhaps he assumed that they would try to Jake, and he wished to exert an influence of uniformity on their technique. Perhaps he was running his own private "experiment" on jaking versus ruling. Possibly, he was merely being thorough in terms of technique. In any case, had he not made such a seemingly mundane choice - two minutes of instruction - the themes I describe in this chapter would never have emerged. Such is the nature of ethnographic research!

The emergent theme that I have discussed includes technology dependence, the obsolescence of nontechnological field skills, and a lack of understanding of the difference between precision and accuracy. It is worth noting that three other unrelated themes emerged from my ethnographic immersion in this field camp class. These other themes are beyond the scope of this chapter, and form the basis for future work. I include them here to contextualise this chapter within the broader ethnographic study I conducted.

The first unrelated theme has to do with risk behaviour. I noted within the main culture-sharing group a division between those students who were risk-averse and those who were not when it came to movement over the landscape. Risk-averse students moved methodically over the terrain so as to minimise their contact with steep slopes, sheer drops and slippery surfaces. Other students appeared alternately to ignore or to seek out risks, moving across the landscape with minimal concern for exertion or even for their own safety. The interaction or "bumping" between these two groups had implications for each group's learning. The second unrelated theme concerns student self -perceptions of athlcticism and physical fitness with regard to their success in the class. Many students perceived physical fitness as a gatekeeper of their final grade, as well as their intrinsic understanding of geological processes. The third unrelated theme builds on the previous two listed. It concerns how teaching assistants acted as gatekeepers to learning by how they addressed (or chose not to address) students' questions and difficulties with physical exertion and the local geology. This gatekeeper behaviour was often related to perceived physical fitness and risk behaviour.

Returning to the technology-dependence theme and its components described in this chapter, the next question to be addressed is, what is to be made of them?

Interpreting the Student Experience

The question, "What is to be made of them?" is the one that Wolcott recommends for discussion in the "Interpretation" section of an ethnographic study. By "them" Wolcott means the members of the culturesharing group, but I expand that to include not only the students I observed, but also the emergent themes and what those themes imply. What is to be made of the students who are technology dependent, who do not value doing things by sight or by hand, who do not know the difference between accuracy and precision?

What is to be made of the fact that they are technology dependent? If field camp is a key benchmark in the transition from student to scientist, and from novice to expert, then we must expect students to internalise, and then to master, a particular skill set. This at least includes locating oneself, judging spatial relationships, and the use of appropriate tools (technology) in the field. If students are dependent on technology to locate themselves and judge spatial relationships, then they have not internalised these skills. Externalised skills are not mastered skills. Students may, at best, become facile with tools such as GPS and Jacob staves. However, they remain near the novice end of the continuum if they do not master spatial skills such as locating themselves using the landscape and ruling section. They must also develop their own affective confidence in these abilities, which facilitates mastery.

What of student dismissal of "old-school" skills? As scientists, mentors and teachers, we should probably look inward to answer this question. How do we present these "by sight/by hand" skills? How do we appear to value them? Are we more like the stereotypical crusty old professor, grunting through clenched teeth, "You lousy kids! Back in my day we had to do this by hand! And we liked it!" Or do we teach these skills with the purpose of building affective and cognitive mastery of field geology? To say to the students, "Because you may not have a GPS/' as justification for "oldschooling," is not sufficient. The students simply do not see themselves as ever being in a position where they do not have access to technology. More importantly, as Joey-Ramon stated, students tend not to question technology. It simply does not occur to them to compare an instrument's reading with their own internal skills, judgment or calibration. They do not ask, "What is this instrument telling me? Does this correspond with what I am actually seeing? What do these numbers mean?" This is what prevented them from realising that they were holding their staves oblique to dip, and not perpendicular to dip.

Finally, what of the students not appreciating the difference between accuracy and precision? The primary issue here is that this informs their misunderstanding of how science works. They equate high levels of technology with high levels of truth, both of which have an ostensible inverse relationship with the human processes of observation and judgment. At worst, this suggests a cadre of future geoscientists who accept as reality whatever their instruments tell them. More likely, it suggests that this cadre will struggle to resolve discrepancies between their high-technology and low technology data sources. It also suggests that they will lack confidence in their internal skills of appraising and understanding a field setting.

The accuracy/ precision issue further suggests that as part of their training as scientists, students should be taught to appraise a situation to apply an appropriate level of precision (technology), rather than to assume that more is better. If students (and scientists) equate precision with accuracy, and accuracy with scientific reality, then they run the risks of misapplying their tools, and being precisely wrong. These risks are compounded by the fact that students will not be able to self -check to avoid those risks, because they do not perceive the problem to begin with.

The good news is that the accuracy /precision issue provides ample opportunities for students to explore the complexities of physical and scientific reality. By developing a clear understanding of these concepts, students can 1) develop greater insight into the workings of science; 2) understand that in some settings, low levels of technology are perfectly appropriate, 3) internalise nontechnological skill sets; 4) develop confidence in their skills; and 5) move ever closer towards the "expert" endpoint on the continuum that marks their journey from student to scientist. Joey-Ramon represents what is known as a "digital native," while Professor X does not. These are conflicting conceptions of technology. Future research could explore how these different concepts of technology shape the expectations that students and faculty have of fieldwork. Do conflicts exist? Do students - and administrators - see field camp as less and less relevant in the face of rapidly developing technology? How does technology impinge upon field education pedagogy?

An Ethnographer Considers Field Camp

Field camp represents a late opportunity to intervene in the training of geoscientists in order to 1) fully internalise their field skill set; 2) increase their confidence in this skill set; 3) explore the complex nature of scientific reality; and 4) provide opportunities for reflection and synthesis. The reduction in the number of field camp classes nationally corresponds to a reduction in the quantity and quality of the training of future geoscientists.

It was clear to me that the students I observed had limited experience in certain spatial skills (locating themselves, ruling section), and little confidence in their cognitive abilities to do so. It was also clear that they had an unsophisticated approach to the application of technology, assuming that in every situation, more is better. They also felt that non-technological strategies are obsolete. These students were all advanced undergraduates, ready to graduate within a year.

Pavlis, et al., (2010) argue that paper-based mapping is obsolete, and must make way for GIS technology. They argue that students must be technologically current with industry and research institutions. I do agree that students should be facile with current technology, for the benefit of both the science and their own careers. However, my position is that for novice students, by-sight/ by-hand skills are critical to their geological praxis, and that the technological applications described by Pavlis, et al., are appropriately applied only once these basic skills are mastered.

Field camp, then, is likely the last chance in the student journey in which we have influence to grow them into the scientists we want them to be. Where else is there a natural laboratory to explore the differences between precision and accuracy? What better setting to explore the efficacies of different levels of technology? What better place to develop, through application and practice, students' spatial abilities, independent of technology? What better outcome than for them to leave the field confident in their newly acquired skills? It is not enough to put students in front of rocks; they have to be put in front of a scientific situation, with the attendant challenges, pitfalls, unknowns and risks that exist in "real" scientific problems. Field camp is not just an expensive outdoors-oriented class; it is the opportunity to grapple and struggle with the real problems and issues in science. Unlike classroom settings, field camp has built-in opportunities for reflection: during mealtimes, around the campfire and in transit.

Field camp forces students to think about a problem even when they are not directly facing the problem. I wondered, and I still wonder, how the students would have felt about their jaking if they could have seen themselves doing it. Would they notice their misalignment with dip? Would it matter to them if they did? What would they do with the opportunity to reflect? It is my hope that this ethnographic study might prompt field camp instructors to integrate more direct reflection opportunities for students to consider the issues of technology dependence, "old-school" skills and the nature of field science. Reflection and synthesis are high orders of learning, and supposedly those that we value among our scientists. Perhaps successful field students are given and take- ample opportunities to engage in reflection and synthesis.

Riggs, Lieder and Balliet (2009) indicated that they could not use their quantitative data and analyses to explain how students make the best field-navigation choices. Perhaps their students with good traverses took ample opportunities to reflect on their observations and behaviours. Perhaps those students did not get "bogged down" in misusing technology. Perhaps they did not have an issue with technology dependence. Perhaps it was their sophisticated understanding of scientific reality that allowed them to test multiple hypotheses as they traversed the field area. All these factors would have aided their mastery of and confidence in their nontechnological field skills. In that case, they reaped the full benefits of the field-learning environment. That environment has no substitute.

Ethnographic study of field camp is an approach to understanding the essence of students' experiences as they learn in the novel spaces of the field. Meaningful themes emerge from the documentation and analysis of behaviours, choices, verbal and nonverbal interactions, and both formal and conversational interviews. In my study of a field camp class, I observed students struggle to measure a stratigraphie section. They rejected an appropriate, low level of technology in favour of a higher level that they misapplied and failed to master. They expressed a strong dependence on this technology. These students systematically devalued low-technology "by hand/by sight" skills such as locating themselves using the landscape and ruling section. They equated progressively higher levels of technology as being equivalent to progressively higher levels of "accuracy," which they consistently confused with "precision." Finally, they equated their misconception of "accuracy" with scientific reality.

Field camp is likely the last chance we have as educators to shape the training of future geoscientists. Field camp is also the best natural laboratory for students to explore affective, semiotic and meta-cognitive factors that shape not only their grade in the class, but also shape their essence as scientists. As field camp goes, so goes our discipline.

Knowledge about P-waves Generated during Particle Motion

Earthquake Teaching and Learning in Thailand

Science teaching and learning in Thailand is geared towards helping students acquire knowledge and skills on their own, with the purpose of encouraging students to think like scientists about authentic problems. However, by the last decade some basic science concepts, such as earthquakes, were not contained in the national science curriculum standards for Thai elementary to secondary school students. Earthquakes have been one of the basic science concepts that all secondary school students in Thailand are expected to learn starting in 2001.

The main sub-topics for an earthquake concept are the definitions of an earthquake, focus, and epicentre, characteristics of earthquakes, causes of earthquakes, seismometres, seismic waves, Richter magnitude scale and Mercalli intensity scale, safety procedures from earthquakes, history of earthquakes in Thailand, active faults and risk areas of earthquakes in Thailand, and Tsunamis and safety procedures from Tsunamis. Therefore, this earthquake concept is not only a new concept for the teaching experience of some Thai teachers, but also a new concept for Thai secondary school students. That inspires the authors to consider in the light of helping Thai students learn about this new concept.

Prior Knowledge

Ausubel's (1968) well known maxim has stated that "The most important single factor influencing learning is what the learner already knows. Ascertain this and teach him accordingly". This concerns the role of existing knowledge in the light of constructivist philosophical frameworks of learning theories. Furthermore, it is evident that what students already know is the primary resource for teaching strategies. Consequently, to help Thai students learn about the earthquake topic, our preliminary study focuses on what students already know about earthquakes.

Generally, students' prior knowledge has originated from the everyday experiences of students. Prior knowledge was first documented by Piaget in 1929, however systematic research into prior knowledge really began in early 1970s. Research into students' prior knowledge has a long history. Previously, coupled with the term "prior knowledge" are the broad labels "preconception", "children's science", "naive knowledge", "children's belief", "intuitive idea" as well as other names. To avoid confusion with differing definitions for these terms, we choose the generic term "prior knowledge" to represent conceptions held by our study population. The students' prior knowledge provides an indication of the alternative conceptions as well as the scientific conceptions possessed by the students.

Implications of Prior Knowledge to Instruction

Prior knowledge plays a major role in the initial stage of conceptual change, and influences how and what students learn. The implications of prior knowledge in instructions depend on the theoretical approaches to conceptual change that instructors use. Until now, there are two major theoretical approaches to conceptual change evidenced by two groups of conceptual change researchers that are considered (1) knowledge-as-theory and (2) knowledge-as-elements. In the same meaning, Elby (2000) viewed these two perspectives from the point of a constructivist and called them (1) misconceptions constructivists and (2) fine-grained constructivists.

The Knowledge-as-theory Perspective (Misconceptions Constructivists)

These researchers considered structural properties of students' knowledge as a coherent unified framework of theory-like character. Prior knowledge is organised into theory, schema, or frame forms. This perspective came from Piagetian learning theory. Researchers represented conceptual change approach by using Piaget's concepts of assimilation and accommodation and Kuhn's (1962 concepts of normal science and scientific revolution. This was defined by Posner et al. in 1982. To change students' prior knowledge in instructions, students' alternative conceptions were used as instruments to construct cognitive conflicting situations between old and new knowledge. If students are dissatisfied with prior knowledge, and find that a new conception is intelligible, appears initially plausible and fruitful, students' conceptual change is likely to occur. Students' alternative conceptions will be replaced by scientific conceptions. Since a prior knowledge structure is like a theory or a model, it can be replaced by other models in the appropriate situations of a cognitive conflict.

Historically, the knowledge-as-theory perspective was a well known perspective suggesting the prior knowledge structure and the conceptual change strategy. Many conceptual change researchers agree with this idea. However, more recently (1990s) some researchers are questioning the nature of the students' prior knowledge structure, especially the coherence of prior knowledge. These researchers suggest that student reasoning is fractured and not as theory-based, robust, coherent, and/ or stable across time and contextual as some theories prescribe. Since prior knowledge is not independent from the cognitive artifacts within a learners' conceptual ecology, it is highly resistant to change. Interestingly, this is also consistent with Strike and Posner's (1992) suggestion, which stated that "These epistemological assumptions suggest that the basic problem of understanding cognitive development is to understand how the components of an individual's conceptual ecology interact and develop and how the conceptual ecology interacts with experience".

The Knowledge-as-elements Perspective (Fine-grained Constructivists)

In contrast with the knowledge-as-theory perspective, this group of researchers considered students' knowledge structure as quasi-independent elements. Prior knowledge is a collection of quasi-independent simple elements within a larger conceptual ecology that are loosely connected into larger conceptual networks without an overarching structure. This perspective came from diSessa's (1993) idea called "knowledge in pieces". diSessa (1993) suggested 4 parts of a knowledge system: (1) elements- describe the size and character of the knowledge structures involved, (2) cognitive mechanism- provides an image of the operation of the prior knowledge system, (3) development understanding the genesis and development of the system, and (4) systemticity- describes the level and kind of relatedness of the elements in the system. diSessa (1993) defined a term "phenomenological primitive" (or p-prim for short) to describe a small knowledge structure (an element). Moreover, diSessa (1993) explained possible characters of p-prims such as, self-explanatory - something happens "because that's the way things are". Pprims constitute a rich vocabulary through which people remember and interpret their experience. P-prims are consistent over time for individual contexts, but have high contextual sensitivity.

Prior knowledge is a collection of elements (p-prims), which are loosely connected, so adding new elements, reorganising connection, and/ or modifying current elements through an evolutionary process will promote conceptual change. In other words, the underlying idea for constructing conceptual change strategies is the finegrained analysis of student reasoning. Since a prior knowledge structure is like loosely connected elements, its elements can be reorganised in the connection by using several appropriate contexts, in order to build more complex and stable formal knowledge (scientific conception).

Although these two perspectives consider prior knowledge structures and conceptual change approaches in different ways, they consistently agree that students' prior knowledge come from students' daily experiences, and it influences students' learning. Thus prior knowledge is well established as the primary source to promote conceptual change strategy as aforementioned.

How to Survey Students' Prior Knowledge

Studies to identify and analyse student prior knowledge have most commonly utilised interviews, concept mapping, and open-ended tests or multiplechoice tests. Each approach has different advantages and disadvantages.

Generally, multiple-choice tests are widely used as quantitative assessments for investigations of large courses or populations. Qualitative methods, like interviews and open-ended tests, are used to elicit richer information about how students think.

These can be timeconsuming to collect, interpret and use, but yield particularly valuable data. Often, qualitative data are used in the development of quantitative tools. For our preliminary study, we aim to explore students' prior knowledge about a previously unstudied topic, prompting use of open-ended tests and interviews.

Geoscience Research into Students' Prior Knowledge

During the last decades, much research into students' prior knowledge has been conducted in various disciplines of science, including geosciences. Research into students' conceptions in the discipline of geosciences has gradually grown, resulting in significant work across a wide range of concepts. For example, Earth's structure, Earth's materials such as rocks, and minerals, Earth's processes such as mountains, volcanoes, earthquakes, weathering, and erosion, and the geological time have all been investigated in significant detail. However, student conceptions of earthquakes and related phenomena have been investigated in only a small number of studies.

Earthquake Facts and Terms in Our Study

Earthquakes are vibrations within the Earth caused by the rupture and sudden movement of the lithosphere - the outer, rigid shell of the Earth above the asthenosphere. The study of earthquakes is important for scientific, social and economic reasons. Earthquakes attest to the fact that dynamic forces are operating within the Earth. They also provide crucial data about the deep interior of the Earth because seismic waves are changed as they travel through the Earth. Interpretation of the thickness, structure and composition of the crust, mantle and core can be made from the types and speeds of seismic waves moving past each zone. The main types of seismic waves are (1) body waves that can be either primary waves (P-waves) or secondary waves (S-waves), and (2) surface waves that can be either Rayleigh waves or Love waves. When earthquakes occur, P-waves propagate as particles in media vibrate parallel to the direction of wave energy (compression), and transfer seismic energy from the focus along the ray path. P-waves move faster than other waves (called the first earthquake waves), providing a non-destructive sign of earthquake activity, with the potential for more destructive seismic waves to follow.

Students' Prior Knowledge about Earthquakes

Research into students' prior knowledge about earthquakes has identified several key significant alternative conceptions that may impede learning. For example, Leather (1987) found that over 50% of the 11 and 14 year olds UK students in his study believed that earthquakes occur only in hot countries and those earthquakes are caused by heat. This is consistent with the results from Sharpe et al.

(1995), which conducted informal interviews with 9 to 10 year old students in Devon, England about causes of earthquakes and found that these students believed that earthquakes occur mostly in hot countries and are caused by heat inside the earth.

Overall, students of all ages hold ideas about earthquake causes that are quite different from the consensus view of scientists; that is, that most large earthquakes occur near tectonic plate boundaries. Furthermore, Ross and Shuell (1993) found that students have trouble understanding the natural cause of earthquakes.

Very few students considered plate movement to be a cause of earthquakes. They also confused unrelated concepts, including confusion between earthquakes, other natural disasters, and weather conditions. Similarly, although Libarkin et al. (2005) found that most college students can write the term "plate tectonics or fault" when describing the cause of earthquakes; students were unable to explain the meaning of these terms when probed by interviews. In addition, Whitney et al. (2004) found that some students erroneously believed that earthquake occurrence can be easily predicted by unusual animal behaviour or changes in weather.

In addition to earthquake causes, previous research has identified students' prior knowledge about earthquake locations, magnitude and intensity. For example, Oberhofer (1991) found that students commonly believed that a change of one magnitude on the Richter scale corresponds to a difference in released energy of 10 times (e.g., a difference of two magnitudes would be 100 times more energy). These students focused on orders of magnitude, instead of seismic energy; it is unclear whether students recognised a difference between magnitude and energy. Moreover, Oliver and Hannafin (2001) found that students often have trouble in earthquake engineering classes because these students did not know how seismic waves travel through the ground. Interestingly, Kortz et al. (2008) found that after finishing their Lecture Tutorial class, students exhibited the lowest absolute gains on earthquake intensity and magnitude topics evaluated by the Geoscience Concept Inventory (GCI). In contrast, these same students showed moderate absolute gains on the topic of earthquake location and other topics of geosciences.

To help Thai students learn about the earthquake topic, in particular, as a first step in investigating the essential underlying cause of student difficulty with seismic wave motion, our study focuses on the prior knowledge of P-waves related particle motion. Accordingly, the objective of this study is to identify Thai students' prior knowledge about P-waves generated particle motion. Open-ended questions designed by the authors and interviews were used to identify students' prior knowledge. Moreover, we use these results to suggest implications for the design of instructional strategies based on what students bring to the classroom, both in the perspective of knowledge-as-theory and knowledge-as-elements.

Instrumentation: Developing the Open-ended Conceptual Questions

To identify Thai students' prior knowledge about the particle motion during P-wave propagation, open-ended conceptual questions were constructed. These questions were measured for content validity by using the Rovinelli and Hambleton's (1977) formula called Index of Item- Objective Congruence. Five experts, who had at least five years experience in investigating or teaching about seismic waves, were invited to judge the item objective congruence. We constructed a table for each expert to use during the item validation. Each expert assessed the agreement of each item with the stated purpose for the item, and marked: agree (+1 point), in which the item and its purpose correlate, not sure (0 point), or disagree (-1 point), in which the item and its purpose do not correlate. We averaged the scores from all experts for a given item. Rovinelli and Hambleton (1977) have suggested for the guidelines for interpretation that the number of content experts be considered when determining a criterion for item acceptance. In this situation in which five content experts are being used to assess a set of items, a minimal criterion might be the index value that would be attained if a minimum of four of five experts classified an item as a perfect match to an objective (+1 point), while one of five are not able to make a decision (0 point). Clearly, if five experts are used, a value of approximately 0.80 might be used as an accepted value. However, Turner and Carlson (2003) suggested that although the cutoff value is a floating criterion, a generally accepted value might be a minimum of 0.75.

Thus our items with Index of Item-Objective Congruence "65 0.80, equivalent to overall agreement that the item matches its stated objective, were selected for inclusion on the questions. We also modified our questions based upon expert suggestions. Finally, these pilot questions were administered to 71 pilot secondary school students who have studied an earthquake topic (not included in the study population). To check students' interpretations of items that matched researcher intentions, we then analysed these students' responses and found that all of those matched the item objective (not considered scientific or alternative conceptions). Question 1(Ql) and question 2 (Q2) were used to ask all study population (n=171) to draw and write about particle motion at P-wave arrival. Question 3 (Q3) was used to interview some study population (n=38) in the same concept.

Context: University Students in Thailand

All participants were Thai freshmen (mean age 19 students enrolled in science, engineering, agricultural sciences, and medicine at a university in the northeast part of Thailand. Although these students had not taken geosciences in the university, they had studied the P-wave concept in prior secondary school instruction. The instruction took various forms, including traditional lecture, lecture with demonstration via internet animations, cookbook experiments, student presentations, and problem-solving. Finally, all students in this study had taken a first-semester introductory physics course that included the topic mechanical waves.

Procedure

In the first step, this study employed two open-ended conceptual questions (Q & Q2), completed by 171 students (about 57% female). Participants were asked to complete the "earthquake conceptual survey" that included these two questions, in addition to eight other open-ended or multiple-choice earthquake questions unrelated to this study, at the beginning of Physics Laboratory at the middle of second semester. In general, students took 10-15 minutes to complete this survey.

Both Q1 and Q2 were used to characterise students' ideas about particle motion at P- wave arrival. We found both similarities and differences in the shape of students' drawings in Ql, thus we grouped them into a structurally similar categories group. For the group of drawings we were unable to category, we invited those students to explain the meaning of their drawings. In the same way, we randomly selected a few students in each structurally similar categories group to explain the meaning of their drawings. Ultimately, there were thirty-eight students (about 60% female), who were invited to the individual interviews. After these students explained their drawings' meanings in Q1, they were given Q3 to confirm their responses again. The individual interview took 3-4 minutes per student and the researcher took note of students' responses. In brief, the final number of participating students in this study was 171 students (38 students in the interview process came from the same population).

Analysis

The collected data, including drawings from Ql, explanations from Q2, and interview responses from Q3, were analysed by grouping main ideas. This involves classifying students' drawings from Ql, identifying keywords in students' explanations from Q2 and grouping students' responses in the individual interviews from Q3. This analysis was measured the intercoder reliability by the authors by using the Holsti (1969) method. It revealed more than 80% of the intercoder reliability between the authors. This means that the authors were in agreement more than 80% of the data analysis. In particular, data were analysed by:

Step 1: Reading students' drawings (Ql) or explanations (Q2) carefully to identify their main ideas.

Step 2: Grouping the same structural drawings from Q1 in a given category.

Step 3: Inviting a few students in each category to individual interview, coupled with all students who drew different structural drawings in Ql,

Step 4: Interviewing all students in Step 3 by using Q3.

Step 5: Splitting scientific and alternative conceptions from Ql, Q2, and Q3's responses.

Step 6: Paraphrasing and classifying alternative responses from Ql, Q2, and Q3.

Step 7: Measuring the intercoder reliability for the paraphrasing and classifying in Step6.

Step 8: Grouping similar alternative responses into a category.

Step 9: Analysing each alternative responses category from Ql, Q2 and Q3, including the scientific group

This study identifies common ideas held by Thai freshmen about particle motion at P-wave arrival as probed by Ql, Q2, and Q3. In this section, we discuss each group of students' drawings from Ql and students' explanation from Q2, as well as descriptions in more details. Moreover, we show the agreement of interview responses with categories of Ql and Q2.

Drawings of Particle Motion (Q1)

Students (n =171) were asked to provide a drawing to expose their ideas about particle motion at P-wave arrival. We identified a range of prior knowledge from students' responses to this question, and found the 5 categories of students' drawings. These 5 categories of the drawings represented different prior knowledge that are (1) the belief that particles spread in all directions at P-wave arrival, (2) the belief that particles move forward like a sine wave motion at P- wave arrival, (3) the scientific idea that particles horizontally move back and forth at P-wave arrival, (4) the belief that particles vertically move back and forth at P- wave arrival, and (5) other beliefs such as particles move forward, or particles go up when P-waves arrive.

Nearly half of these students believed that particles will spread in all directions when affected by P-waves. These students represented the idea in several different ways. The most common drawing was that of a series of concentric circles surrounding a central dot, followed by a drawing of arrows spreading in all directions from a common origin. A third, less common drawing mixed concentric circles and arrows. Interview responses indicate that all three drawing types were indicative of the same alternative conception. For example, students explained each of these drawings, respectively, by stating, "the particle spreads in all directions at the first earthquake wave arrival like water waves when I throw the rock down", "the particle moves in all directions", and "the particle spreads in all directions".

This was an example dialogue between the researcher and a student in the individual interview. The researcher asked a student to describe the meaning of his drawing and followed-up with giving Q3. Eventually, the researcher was able to both identify a student's drawing and know the alternative conception about particle motion at P-wave arrival. A student's drawing in the following case was grouped into Group 1.

Explanations of Particle Motion (Q2)

Students (n=171) were asked to write an explanation about particle motion at P- wave arrival. These explanations were paraphrased, intercoder reliability tested, and grouped into five main ideas. This shows the five groups with English titles, the percentages of students in each explanation group, and an example from each group in both the original Thai and in an English translation.

Overall, students' explanations were easily grouped into the five main categories. For example, "the particle will vibrate along the direction of the arrow shown in the figure", "the particle horizontally vibrates like sound waves", and "the particle vibrates in the same direction of the wave propagation" were clustered into the same idea, which was the scientific conception. These explanations were paraphrased to group by their authentic meaning.

However, some students' explanations were unable to group because of obscurity, irrelevance and meaningless. They were mixed in Group 5. For example, "the particle vibrates with high frequency", "the particle quakes some buildings and collapses them", "the particle vibration depends on the type of the medium." were explanations in Group 5. The most popular explanation was that at P-wave arrival particles move with a sine wave motion, similar to the seismogram shown in Q2. The students' responses to Q3 are the same with both Ql and Q2. We found that 36% of these students believed that P-waves will shake the particle in a medium like that of water waves. The others 29%, 16%, and 10% of them believed that P-waves will shake the particle in a medium like that of waves on a string, sound waves, and light respectively.

Our study utilised open-ended questions constructed by the authors to identify Thai university students' prior knowledge about particle motion at P-wave arrival. Ql and Q2 were administered to 171 students, as well as from the same study population we invited 38 students for the individual interview by using Q3.

We paraphrased and classified students' drawing in Ql, their explanations in Q2, and interview responses in Q3. We ultimately grouped their ideas into five common categories for both Ql and Q2 based on strong confirmation from interviews.

Implications of Instruction

Our study revealed Thai students prior knowledge about particle motion at P-wave arrival. Prior knowledge is considered to be a basic source for the conceptual change approaches for both the knowledge-as-theory perspective and the knowledge-as-elements perspective as aforementioned. Since the views of the nature of structural prior knowledge of these two perspectives are different, the instructional designs for classroom conceptual change strategies of them are also different.

In the light of knowledge-as-theory, the main goal of a conceptual strategy is a cognitive conflict. Instructors have to engage students to confront their prior knowledge. In other words, instructors should provide students with scenarios or questions that cannot be explained with common alternative conceptions. In this situation, students will be dissatisfied and learn why their current ideas must be abandoned, often deciding to abandon these ideas on their own. Ultimately, that alternative conception will be replaced by the scientific conception.

Although constructivism views students' prior knowledge as a primary source for learning, erasing alternative conceptions with a replacement of the scientific conception is inconsistent with constructivism. Smith et al. in 1993 stated that "Our central claim is that many of the assertions of misconceptions research are inconsistent with constructivism. Misconceptions research has emphasized the flawed results of student learning. Constructivism, in contrast, characterises the process of learning as the gradual recrafting of existing knowledge that, despite many intermediate difficulties, is eventually successful. It is difficult to see how misconceptions that (a) interfere with learning, (b) must be replaced, and (c) resist instruction can also play the role of useful prior knowledge that supports students' learning. If we take constructivism seriously, we must either reconsider the solely mistaken character of misconceptions or look for other ideas to serve as productive resources for student learning". However, from this theoretical debate Elby (2000) distinguished constructivists into two flavours such as misconceptions constructivists, and finegrained constructivists.

So in the light of knowledge-as-elements, to promote a conceptual change approach instructors should focus on how prior knowledge elements are activated in appropriate contexts. For the same concept instructors should encourage students to confront with various contexts. The fine-grained analysis of student reasoning suggests that conceptual change requires restructuring, editing, and reorganising the loosely connection of prior knowledge elements. In our study we found students' prior knowledge about particle motion at P-wave arrival, which is likely to change in different contexts. This revealed not only a highly contextual sensitivity property of elements but also pointed out a pathway to bridge such loosely linked elements in order to be concrete elements of a scientific conception. Students should experience in multiple contexts to reorganise their prior knowledge elements.

For example, to promote conceptual change about particle motion at P-wave arrival guided by the prior knowledge in our study, instructors may ask questions in different contexts or show different demonstrations in this concept. Instructors may use a demonstration wave spring, coupled with a red string stuck on its helix to demonstrate the particle motion at P-wave arrival.

Instructors may throw a rock in a pool and ask students to look at the movement of leaves floating on that pool. Instructors may introduce students to play human waves or act like a particle of waves' medium. Moreover, instructors may use a simple seismograph to teach about seismic waves concepts. It is possible that these various contexts may help students to reorganise their loosely connected existing ideas into larger conceptual networks and ultimately go to the network of a scientific conception.

The research presented here suggests that Thai freshmen hold a variety of prior knowledge about particle motion at P-wave arrival. This has originated from their everyday experiences. However, the main categories of the prior knowledge are (1) the belief that particles spread in all directions, like water waves, when P-waves arrive, (2) the belief that particles move forward with a sine wave motion, and that these particles travel with the propagating wave energy to the P-wave's final destination, (3) the belief that particles vertically move back and forth at P- wave arrival. These beliefs are the alternative conception held by more than three-quarters of Thai freshmen in our study.

The other held the scientific conception (category 4) that particles in a medium vibrate in the same direction as the propagating wave energy when P-waves arrive, coupled with recognition that particles do not travel with the propagating energy. Furthermore, these students' prior knowledge elements varied in different question contexts. However, these multi contexts may be used to design a conceptual change strategy which is based on the knowledge-as-elements perspective, in which prior knowledge is a collection of quasi-independent simple elements within a larger conceptual ecology that are loosely connected into larger conceptual networks without an overarching structure. Since scientific ideas require the reorganising and restructuring of the connection of prior knowledge elements, multiple contexts are required. Indeed, this prior knowledge is still the primary source of cognitive conflict to promote a conceptual change strategy in the light of knowledge as theory perspective. In brief, the prior knowledge implications for classroom instruction depend on how instructors think about the nature of structural prior knowledge, including its conceptual change approach.

In future work, the results of this study will be used to construct a teaching module for encouraging conceptual change of Thai secondary school students. This teaching module will contain problems in various contexts guided by this prior knowledge. Earthquake news, online earthquake data from educational websites and real seismograms from local seismometres will be used to engage students in classrooms. Students will perform experiments by using a simple seismometre to learn about seismic wave motion.

Bibliography

Al Marhoun, M.A.: *Pressure-volume-temperature Correlations for Saudi Crude Oils*, Society of Petroleum Engineers Publication, UK, 1985.

Bouma, A., Normark, W. R., and Barnes, N. E.: *Submarine Fans and Related Turbidite Systems*, Springer Verlag, New York, 1986.

Brantly, John Edward: *History of Oil Well Drilling*, Book Division, Gulf Pub. Co., Houston, 1971.

Clark, James Anthony, C. A. Warner, and H. E. Walton: *The Chronological History of the Petroleum and Natural Gas Industries,* Clark Book Co., Houston, 1963.

Conaway, Charles F.: *The Petroleum Industry: A Nontechnical Guide,* PennWell Pub. Co., Tulsa, OK, 1999.

Davies R. J., Cartwright J., Stewart S. A, Underhill, J. R., Lappin M.: *3D Seismic Technology: Application to the Exploration of Sedimentary Basins*, Geological Society, London, Memoir, 2003.

Dunne, T. and L.B. Leopold: *Water in Environmental Planning*, W.H. Freeman and Co., San Francisco, CA, 1978.

Edwards, Denis: *Earth Revealing Earth Healing: Ecology and Christian Theology,* Collegeville: Liturgical Press, 2001.

Egan, Walter G.: *Photometry and Polarization in Remote Sensing,* New York, 1985.

Fenichell, Stephen: *Plastic: The Making of a Synthetic Century,* Harper Business, New York, 1996.

Ford, Trevor D. and Cullingford, C. H. D. (editors), 1976, *The Science of Speleology:* Academic Press, New York, 593 p.

Gray, Forest: *Petroleum Production in Nontechnical Language,* PennWell Books, Tulsa, OK, 1995.

Hallam, A.: *A Revolution in the Earth Sciences:* New York, Oxford University Press, 1973.

Hannum, M.B.: *Late Quaternary Evolution of the Kennebec and Damariscotta River Estuaries*, Geological Sciences, Univ. Maine, Orono, 1997.

Khilyuk, Leonid F., George V. Chilingar, and Herman H. Rieke: *Probability in Petroleum and Environmental Engineering*, Gulf Pub. Co., Houston, 2005.

Kondolf, G. Mathias; Hervé Piégay: *Tools in Fluvial Geomorphology*, Wiley, New York, 2003.

Langran, G.: *Time in Geographic Information Systems,* Taylor & Francis, Bristol, 1992.

Laurence M. Harwood, Christopher J. Moody: *Experimental Organic Chemistry: Principles and Practice*, WileyBlackwell, Oxford, 1989.

Manthorp, P.A.: *Estuarine Circulation and Sediment Transport in the Saco River estuary*, Boston University, Boston, MA, 1995.

Peters, K.E., and J.M. Moldowan: *The Biomarker Guide: Interpreting Molecular Fossils in Petroleum and Ancient Sediments*, Prentice-Hall, Inc., Englewood Cliffs, 1993.

Peters, K.E., C.C. Walters, and J.M. Moldowan: *The Biomarker Guide,* Cambridge University Press, New York, 2005.

Peuquet, D.: *Representations of Space and Time.* New York: Guilford Press, 2002.

Ritter, D.F., Kochel, R.C., and Miller, J.R.: *Process Geomorphology*, Wm.C. Brown Publishers, Dubuque, Iowa, 1995.

Robinson, I. S.: *Satellite Oceanography: An Introduction for Oceanographers and Remote-sensing Scientists*, New York, Halsted Press, 1985.

Scott, A.C.: *Coal and Coal-bearing Strata: Recent Advances*, Blackwell Scientific Publications, Geological Society Special Publication, Boston, 1987.

Selby, Michael John: *Earth's Changing Surface: an Introduction to Geomorphology*, Clarendon Press, Oxford, 1985.

Thomas, L.: *Handbook of Practical Coal Geology*: John Wiley & Sons, Chichester, 1992.

Tissot, B., and D.H. Welte: *Petroleum Formation and Occurrence*, Springer-Verlag, New York, 1984.

Wainwright, John; Mulligan, M.: *Environmental Modelling: Finding Simplicity in Complexity*, John Wiley and Sons Ltd, London, 2003.

Wolf, P. R.: *Elements of Photogrammetry,* New York, 1983.

Yang, Y., R. Zou, Z. Shi, and R. Jiang: *Atlas for Coal Petrography of China*, China University of Mining and Technology Press, Beijing, 1996.

Zim, Herbert S. and Paul R. Shaffer: *Rocks and Minerals: A Guide to Familiar Mnierals, Gems, Ores and Rocks,* Golden, New York, 1957.

Index

❑❑❑